Hydrogels in Cell-Based Therapies

RSC Soft Matter Series

Series Editors:
Hans-Jürgen Butt, *Max Planck Institute for Polymer Research, Germany*
Ian W. Hamley, *University of Reading, UK*
Howard A. Stone, *Princeton University, USA*
Chi Wu, *The Chinese University of Hong Kong, China*

Titles in this Series:
1: Functional Molecular Gels
2: Hydrogels in Cell-Based Therapies

How to obtain future titles on publication:
A standing order plan is available for this series. A standing order will bring delivery of each new volume immediately on publication.

For further information please contact:
Book Sales Department, Royal Society of Chemistry, Thomas Graham House, Science Park, Milton Road, Cambridge, CB4 0WF, UK
Telephone: +44 (0)1223 420066, Fax: +44 (0)1223 420247
Email: booksales@rsc.org
Visit our website at http://www.rsc.org/books

Hydrogels in Cell-Based Therapies

Edited by

Che J. Connon
School of Pharmacy, University of Reading, UK
Email: C.J.Connon@reading.ac.uk

and

Ian W. Hamley
Department of Chemistry, University of Reading, UK
Email: i.w.hamley@reading.ac.uk

RSC Soft Matter No. 2

ISBN: 978-1-84973-798-2
ISSN: 2048-7681

A catalogue record for this book is available from the British Library

© The Royal Society of Chemistry 2014

All rights reserved

Apart from fair dealing for the purposes of research for non-commercial purposes or for private study, criticism or review, as permitted under the Copyright, Designs and Patents Act 1988 and the Copyright and Related Rights Regulations 2003, this publication may not be reproduced, stored or transmitted, in any form or by any means, without the prior permission in writing of The Royal Society of Chemistry or the copyright owner, or in the case of reproduction in accordance with the terms of licences issued by the Copyright Licensing Agency in the UK, or in accordance with the terms of the licences issued by the appropriate Reproduction Rights Organization outside the UK. -Enquiries concerning reproduction outside the terms stated here should be sent to The Royal Society of Chemistry at the address printed on this page.

The RSC is not responsible for individual opinions expressed in this work.

Published by The Royal Society of Chemistry,
Thomas Graham House, Science Park, Milton Road,
Cambridge CB4 0WF, UK

Registered Charity Number 207890

For further information see our web site at http://www.rsc.org

Contents

Chapter 1 Soluble Molecule Transport Within Synthetic Hydrogels in Comparison to the Native Extracellular Matrix   1
Matthew Parlato and William Murphy

    1.1 Introduction   1
    1.2 Steady-State Diffusion   2
        1.2.1 Steady-State Diffusion Within the Native ECM   2
        1.2.2 Steady-State Diffusion Within Synthetic Hydrogels   5
    1.3 Soluble Factor Generation and Consumption   8
        1.3.1 Soluble Factor Generation and Consumption Within the Native ECM   8
        1.3.2 Soluble Factor Generation and Consumption Within Synthetic Hydrogels   11
    1.4 Matrix Interactions   13
        1.4.1 Interactions with ECM Components   13
        1.4.2 Interactions with Hydrogel Components   14
    1.5 Temporal Dependencies   17
        1.5.1 Temporal Dependencies During *In Vivo* Transport   17
        1.5.2 Temporal Dependencies During Transport in Synthetic Hydrogels   20
    1.6 Convection   24
        1.6.1 *In Vivo* Transport by Convection   24
        1.6.2 Transport in Synthetic Hydrogels by Convection   25
    1.7 Future Directions and Concluding Remarks   26
    References   27

Chapter 2	**Biocompatibility of Hydrogelators Based on Small Peptide Derivatives** *Yi Kuang, Ning Zhou, and Bing Xu*	31
	2.1 Introduction	31
	2.2 Biocompatibility of Hydrogelators in Mammalian Cells	33
	2.2.1 Biocompatibility of Hydrogelators of Conjugates of Pentapeptides and Aromatic Motifs	33
	2.2.2 Biocompatibility of Hydrogelators of N-Unsubstituted Dipeptides	36
	2.2.3 Biocompatibility of Hydrogelators of the Conjugates of Nucleobases and Peptides	38
	2.2.4 Biocompatibility of Hydrogelators of Conjugates of Nucleobases, Peptides, and Saccharides	40
	2.3 Biocompatibility of Hydrogelators in Animal Models	42
	2.4 Conclusions	45
	References	45
Chapter 3	**Recombinant Protein Hydrogels for Cell Injection and Transplantation** *Patrick L. Benitez and Sarah C. Heilshorn*	48
	3.1 Overview	48
	3.2 Motivation: Cellular Control Though Tailored Protein Interactions	49
	3.2.1 Extracellular Protein Interactions Relevant to Cell Delivery	50
	3.2.2 Domain-Level Engineering in Recombinant Hydrogels	52
	3.3 From Concept to Protein-Engineered Cell Delivery	55
	3.4 Case Studies: Recent Developments and Applications of Recombinant Hydrogels	58
	3.4.1 Elastin	58
	3.4.2 Collagen	61
	3.4.3 Resilin-like Proteins	63
	3.4.4 Mixing-Induced Two-Component Hydrogels (MITCH)	64
	3.5 Challenges and Opportunities in Clinical Translation	66
	3.6 Conclusion	67
	References	68

Contents

Chapter 4	**The Instructive Role of Biomaterials in Cell-Based Therapy and Tissue Engineering**	73
	Roanne R. Jones, Ian W. Hamley, and Che J. Connon	

    4.1 Introduction to Cell-Based Therapies 73
          4.1.1 Potential Clinical Applications and
                Pharmaceutical Industry Involvement 75
          4.1.2 Stem Cell Banking 76
    4.2 Biomaterials 76
          4.2.1 Natural Biomaterials 77
          4.2.2 Synthetic Materials in Tissue Engineering
                Applications 78
    4.3 Biomaterials in Tissue Engineering 80
          4.3.1 Cell Response to Substrate Elasticity 81
          4.3.2 Structuring of ECM Mimics 83
          4.3.3 Mechanosensitivity 84
          4.3.4 Proteins and Pathways 86
          4.3.5 Relevance to Regenerative Medicine 88
    References 89

Chapter 5	**Microencapsulation of Probiotic Bacteria into Alginate Hydrogels**	95
	M. T. Cook, D. Charalampopoulos, and V. V. Khutoryanskiy	

    5.1 Introduction 95
    5.2 The Chemistry of Alginates 97
    5.3 Producing Alginate Hydrogels 99
    5.4 Protecting Probiotics—Demonstrating Efficacy of
         Alginate Microcapsules 103
    5.5 Modifications of Alginate Hydrogels 106
    5.6 The Future of Alginate as an Immobilization Matrix
         for Probiotics 109
    References 110

Chapter 6	**Enzyme-Responsive Hydrogels for Biomedical Applications**	112
	Yousef M. Abul-Haija and Rein V. Ulijn	

    6.1 Introduction 112
          6.1.1 Polymeric and Self-Assembling Hydrogels 113
          6.1.2 Use of Enzymes in Fabrication of
                Next-Generation Biomaterials 113
          6.1.3 Use of Enzymes as 'Stimuli' in
                Smart Materials 114

6.2 Biocatalytic Assembly of Supramolecular Hydrogels		115
6.2.1 Peptide-Based Hydrogels		116
6.2.2 Biocatalytic Peptide Self-Assembly for Biomaterials Fabrication		117
6.3 Biomedical Applications		124
6.3.1 Controlling and Directing Cell Fate		125
6.3.2 Imaging and Biosensing		126
6.3.3 Controlled Drug Release		128
6.3.4 Cell Scaffolds and Tissue Engineering		128
6.4 Conclusions and Outlook		129
References		130

Chapter 7 Alginate Hydrogels for the 3D Culture and Therapeutic Delivery of Cells   135
Bernice Wright and Che J. Connon

7.1 Alginate Isolation and Gelation Chemistry	135
7.1.1 Extraction and Purification of Alginate Polysaccharides	136
7.1.2 Alginate Gelation Chemistry	137
7.1.3 The Alginate Gel Structure	140
7.2 Alginate Hydrogels as Cell Culture Scaffolds	141
7.2.1 Regeneration of the Cornea Using Alginate-Encapsulated Corneal Cells	141
7.2.2 Harnessing the Therapeutic Potential of Embryonic Stem Cells Using Alginate Gels	142
7.2.3 Engineering Clinically Viable Trabecular Bone and Cartilage Using Alginate Gel Scaffolds	144
7.2.4 Alginate Gels for Cardiac Tissue Repair: Development of the Cardiac Patch	148
7.2.5 Alginate/Endothelial Progenitor Cell Platforms for Therapeutic Angiogenesis and Neovascularization	149
7.2.6 The Construction of Neural Prosthetics and Culture Systems Using Alginate Gels	150
7.3 The Influence of Alginate Gel Biophysical and Biochemical Properties on Cell Phenotype	151
7.3.1 The Effect of Alginate Gel Biophysical Properties on Encapsulated Cells	152
7.3.2 The Effect of Biochemically Modified Alginate Gels on Cells	155
7.4 Perspectives	159
References	164

Contents

Chapter 8 Mechanical Characterization of Hydrogels and its Implications for Cellular Activities   171
Samantha L. Wilson, Mark Ahearne, Alicia J. El Haj, and Ying Yang

8.1 Introduction	171
8.2 Hydrogel Characterization Techniques	173
8.3 Effect of Hydrogel Mechanical Properties on Cellular Activities	176
8.4 Effect of Cellular Activity on Hydrogel Properties	178
8.5 Mechanical Properties as a Marker of Cellular Activities	181
8.5.1 Indicator of Differentiation Status	181
8.5.2 Indicator of Cell Viability and Contractility	182
8.5.3 Indicator of Network Structure in the Hydrogel	183
8.6 Strategies for Improving the Mechanical Properties of Hydrogels	184
8.6.1 Concentration	185
8.6.2 Crosslinking	185
8.6.3 Composition	185
8.6.4 Orientation of Fibrous Components	186
8.6.5 Micro- and Nanopatterning	186
8.6.6 Magnetically Aligned Collagen	186
8.6.7 Electrospinning of Nanofibres	187
8.7 Conclusion	187
References	188

Chapter 9 Extracellular Matrix-Like Hydrogels for Applications in Regenerative Medicine   191
Aleksander Skardal

9.1 A Brief Introduction to the Field of Biomaterials	191
9.2 Hydrogel Biomaterial Types	192
9.2.1 Synthetic Polymer Hydrogels	192
9.2.2 Collagen	193
9.2.3 Hyaluronic Acid	193
9.2.4 Alginate	194
9.2.5 Fibrin	194
9.3 Implementations in Regenerative Medicine	194
9.3.1 Stem Cell Culture	195
9.3.2 Primary Cell and Tissue Culture	198
9.3.3 Cell Therapy	200
9.3.4 Tissue Engineering *Ex Vivo/In Vitro*	201
9.4 Future Potential and Implications	208
9.5 Conclusions	209
References	209

Subject Index   216

CHAPTER 1

Soluble Molecule Transport Within Synthetic Hydrogels in Comparison to the Native Extracellular Matrix

MATTHEW PARLATO[a] AND WILLIAM MURPHY*[a,b,c]

[a] Department of Biomedical Engineering, University of Wisconsin Madison, Wisconsin Institutes for Medical Research, 1111 Highland Avenue, Madison, WI 53705, USA; [b] Materials Science Program, University of Wisconsin Madison, Wisconsin Institutes for Medical Research, 1111 Highland Avenue, Madison, WI 53705, USA; [c] Department of Orthopedics and Rehabilitation, University of Wisconsin Madison, Wisconsin Institutes for Medical Research, 1111 Highland Avenue, Madison, WI 53705, USA
*E-mail: wlmurphy@wisc.edu

1.1 Introduction

Soluble factor signalling and gradient formation are of known biological importance and direct processes such as stem cell differentiation,[1,2] cellular migration,[3-6] limb bud development,[7-9] and neural tube development.[7] Soluble transport within the *in vivo* environment is complex, involving spatiotemporal interactions and molecular recognition between soluble molecules and extracellular matrix (ECM) components.[2-16] Because of such complexity, what is known and what can be studied about soluble transport *in vivo* is limited. Therefore, the use of well-defined *in vitro* experimental

RSC Soft Matter No. 2
Hydrogels in Cell-Based Therapies
Edited by Che J. Connon and Ian W. Hamley
© The Royal Society of Chemistry 2014
Published by the Royal Society of Chemistry, www.rsc.org

platforms is an attractive option. Because of the similarity of hydrogels to the native ECM, synthetic hydrogels can serve as model systems for the study of soluble transport and gradient formation within the ECM.[17-19] Synthetic hydrogels are also useful because of their biocompatibility and adaptability for use with a variety of chemistries.[17,19-26]

The hydrated polymer chains of synthetic hydrogels slow solute movement just as the macromolecules within the ECM do, thus assisting in the formation of concentration gradients.[27] Furthermore, drug delivery technologies have been incorporated into synthetic hydrogels that serve as well-defined soluble factor sources and sinks within the hydrogel.[28-31] Other experimental approaches seek to incorporate the ability of the native ECM to specifically bind and release soluble molecules into synthetic hydrogels by the incorporation of proteoglycans[21,26,32-34] or peptides that have high binding affinities for specific soluble molecules.[31,35-37] Many methods also exist that exert temporal control over transport within synthetic hydrogels by allowing the hydrogel to degrade over time, be remodelled by cell-secreted enzymes, or respond to external cues such as temperature or pH.[20,24,30,38-47]

There are many articles and reviews that discuss the first principles of transport within the native ECM and synthetic hydrogels separately;[27,29,48-52] however, the purpose of this chapter is to compare and contrast the two. We endeavour to address some of the critical questions that arise during development of synthetic hydrogels to mimic natural signalling gradients in the ECM, such as: (1) how does transport and gradient formation of soluble molecules within synthetic hydrogels compare to that within the native ECM? (2) What aspects of signalling within the native ECM have been mimicked within synthetic hydrogels and what aspects remain to be explored? and (3) what are the potential consequences of these differences, and how can the synthetic hydrogels be made to more closely mimic the signalling of the native ECM? This chapter is divided into five sections based on the following parameters that influence molecular transport in natural or synthetic ECMs: steady-state diffusion, soluble factor generation and consumption, matrix interactions, temporal dependencies, and convection. Each of these sections is divided into two subsections. The first subsection discusses the topic with regard to the native ECM and the second with regard to synthetic hydrogels. Finally, the chapter concludes with a short discussion of the future directions for synthetic hydrogels that seek to recapitulate various aspects of signalling in the native ECM.

1.2 Steady-State Diffusion

1.2.1 Steady-State Diffusion Within the Native ECM

Soluble factor gradients within the ECM are generated through a variety of mechanisms, but to begin the discussion, a simple case with defined 'source' and 'sink' regions is discussed. In the simplest scenario, defined source and sink regions occur due to one group of cells producing large amounts of soluble molecules while another nearby group does not produce these molecules and instead consumes them. The goal of this section is to understand the basic mechanisms by which soluble factor gradients may form within the native ECM. Additionally, we examine what fundamental properties of both the soluble factor and the native ECM affect this gradient formation.

Within this section, all sources and sinks are assumed to exist at a single point in space to facilitate mathematical descriptions. Therefore, they are referred to as 'point-sources' and 'point-sinks'. They are also assumed to produce or consume molecules instantaneously and without limits. Due to these properties, they are referred to as 'perfect sources and sinks'. Furthermore, we assume that the ECM in which the molecules are diffusing is homogeneous and that all parameters are constant with time (*i.e.* steady state). Notably, biological scenarios do not feature perfect, point-sources and point-sinks, but this is a useful and widely utilized starting point in discussions of transport in the native ECM.[12,15,53] A diagram of this problem is shown in Figure 1.1a. A summary of these assumptions is as follows:

(1) All regions are homogeneous.
(2) The source region is a perfect, point source.
(3) The sink region is a perfect, point sink.
(4) All parameters are constant with time ('steady state').

To analyse this situation, Fick's first law of diffusion is applied (eqn (1.1)) This law states that the mass flux, J, of a solute through a region of space is proportional to the rate of change of the solute's concentration with respect to position, dC/dx. We assume that at the position $x = 0$, the solute is produced in a way that maintains a constant concentration, C_0, and this assumption is used to develop the boundary condition (eqn (1.1), BC 1).

$$J = -D\frac{dC}{dx}$$
$$\text{BC 1}: C(0) = C_0 \tag{1.1}$$
$$C(x) = C_0 - \frac{Jx}{D}$$

Eqn (1.1) shows the diffusion and gradient formation within a homogenous region from Fick's first law. Here C is the concentration of the soluble factor, J is the mass flux of the solute, x is position, and D is the diffusion coefficient. Here a linear relationship is demonstrated that is dependent on initial concentration of the molecules at the source (C_0), mass flux (J), and the solute diffusion coefficient (D). The ratio of J/D is the slope of the linear concentration gradient (Figure 1.1b), and C_0 is the y-intercept. This problem can also be solved by Fick's second law of diffusion. However, this law is applicable with non-steady state problems because it assumes non-constant mass flux, so it is discussed in Section 1.4.1 where temporal dependencies are discussed in detail.

In general, solutes that are small in size when compared to pores in the ECM diffuse quickly, and those that are large diffuse more slowly, because

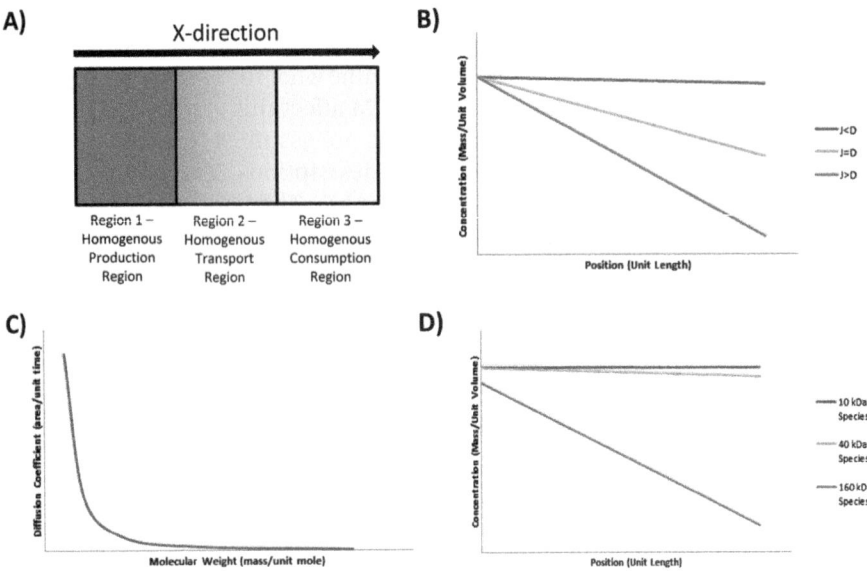

Figure 1.1 (A) Diagram of proposed problem involving a perfect source, perfect sink, and a homogeneous transport region. Arrow indicates the positive x-direction. (B) Analysis of solute mass flux over diffusion coefficient ratio (J over D ratio) for the transport problem diagrammed in (A). When J is far less than D, concentration is nearly constant with respect to x. Conversely, when J is far greater than D, the concentration decreases quickly with respect to x. (C) Diffusion coefficient *versus* molecular weight as proposed by the Brinkman equation for diffusion within the native ECM. A steep, non-linear drop-off of the diffusion coefficient with respect to increasing molecular weight is observed. (D) Changes in the concentration gradient based solely on changes to the diffusion coefficient calculated by the Brinkman equation shown in (C). A fourfold increase in molecular weight only slightly changes the concentration gradient; however, a 16-fold increase in the molecular weight brings about a drastic change in the concentration gradient.

that larger solutes interact with the matrix more often than smaller solutes.[51,54,55] This slowing of molecular movement is referred to as 'sieving action'. Many models exist to approximate the size of a molecule, and a commonly utilized model is the Stokes–Einstein relationship (eqn (1.2)). This model assumes a spherical molecule with a density close to that of water (1 g mL^{-1}).[54] Molecular radii calculated from this relationship are referred to as hydrodynamic radii (represented by the variable a) because this model is based on hydrodynamic diffusion theory.[27] Though derived for a solute diffusing freely in a homogenous solution, this model can also be related to diffusion coefficients of solutes in hydrogels.[27,54,55] One of the best-known models was derived by Brinkman, which relates diffusion coefficients in a gel, such as the native ECM (represented by the variable D_{ECM}), to that in

Soluble Molecule Transport Within Synthetic Hydrogels

water (represented by the variable D_0) based solely on hydrodynamic radius, a, and a fitted system parameter, κ.[54]

$$D = \frac{RT}{6\mu\pi aN}; \text{ where } a = \left(\frac{3MW}{4\pi\rho N}\right)^{1/3}$$

$$\frac{D_{ECM}}{D_0} = \frac{1}{1 + \kappa a + 1/3(\kappa a)^2}$$

(1.2)

Eqn (1.2) shows the Stokes–Einstein relationship for relating the diffusion coefficient, D, to molecular radius, a. In this relationship, an approximately spherical molecule is assumed with a density, ρ, which is close to that of water (1 g mL^{-1}). The notation used here is as follows: R is the universal gas constant, T is temperature, μ is the solution viscosity, N is Avogadro's number, and MW is the molecular weight. Diffusion coefficients within the ECM, D_{ECM}, can be compared to those found in dilute water solutions, D_0, through this relationship developed by Brinkman where κ is a fitted system factor and a is again molecular radius.

Eqn (1.2) presents a method predicting how various biomolecules of different hydrodynamic radii diffuse through the ECM (Figure 1.4c and d). Even when all other variables are held constant, molecular size alone can dramatically affect the slope of the gradient.

Many examples of Fick's first law are found in *in vivo* situations, and studies of the transport of solutes into cancerous tissue provide illustrative examples of such relationships.[56–58] For instance, some types of tumours are known to retain a large amount of fluid, increasing their interstitial volume and thus increasing the diffusion coefficients of solutes.[54] The change occurs because the increased volume lowers the concentration of macromolecules that compose the ECM, thus allowing for faster movement of solutes because there are fewer solute interactions with the matrix. In contrast, within other types of tumours diffusion is severely hindered because of increased ECM protein secretion around the tumour tissue.[56,57] Creation of more material around a site increases the concentration of the macromolecules that comprise the ECM, and transport by diffusion is slowed because of an increased number of interactions between solutes and matrix. Additionally, digestion of the tumour ECM by enzymes enhances diffusion into the tumour tissue, due in part to the corresponding decrease in ECM concentration.[58] From these studies, it is clear that transport within the native ECM is dependent on ECM network parameters in addition to solute properties.

1.2.2 Steady-State Diffusion Within Synthetic Hydrogels

As in the native ECM, gradients can form within synthetic hydrogels through multiple mechanisms; however, numerous examples of gradients forming primarily through sieving action can be found in the literature.[29,42,52,59–61] When analysing gradient formation in synthetic hydrogels, the same assumptions that were made in the previous section are made here. Notably, the assumption that the hydrogel is a homogeneous, porous network of inert macromolecules is more appropriate for synthetic hydrogels than for the native ECM.[17,27,42]

Prediction of the diffusion coefficients of solutes within synthetic hydrogels has been modelled using hydrodynamic theory and the Stokes–Einstein relationship (eqn (1.2)).[27] Specifically, the diffusion coefficient within hydrogels, D_{gel}, versus that of free solution, D_0, is commonly modelled with the following relationship:

$$\frac{D_{gel}}{D_0} = e^{-\left(Bac^{-1/2}\right)} \quad (1.3)$$

Eqn (1.3) shows the Stokes–Einstein relationship specifically for the hydrogel diffusion case. Here a is the hydrodynamic radius, B is a system coefficient, and c is the polymer concentration of the hydrogel (expressed in units of volume fraction). D_{gel} is the diffusion coefficient within the hydrogel and D_0 is the diffusion coefficient of the solute in free solution.[27]

The Stokes–Einstein relationship is a simple representation of the sieving action of a synthetic hydrogel, but this model is limited based on its dependence on a fitted parameter, the system coefficient B. A more broadly useful and predictive model minimizes its dependence on fitted parameters and maximizes its dependence on parameters that can be measured by the experimenter. Based upon thermodynamics first principles of hydrogel networks, the molecular weight between crosslinks, M_c, can be predicted based upon average polymer molecular weight, M_n, and volume fraction of polymer in the network, $v_{2,s}$ (eqn (1.4)):[17]

$$M_c = \left[\frac{2}{M_n} - \frac{(v/V)\left(\ln(1 - v_{2,s}) + v_{2,s} + \chi v_{2,s}^2\right)}{v_{2,r}((v_{2,s}/v_{2,r})^{1/3} - (v_{2,s}/2v_{2,r}))} \right]^{-1} \quad (1.4)$$

Eqn (1.4) shows the molecular weight between crosslinks of a hydrogel based upon thermodynamic first principles. Here M_c is the molecular weight between crosslinks, M_n is the average polymer molecular weight, v is the specific molar volume of the polymer, V is the molar volume of water, χ is the Flory interaction parameter between the polymer and the solvent, $v_{2,s}$ is the swollen volume fraction of polymer, and $v_{2,r}$ is the relaxed volume fraction of the polymer.[17]

Once the molecular weight between crosslinks is known, the average physical distance between crosslinks, ξ, can be determined based on known polymer parameters and the molecular weight between crosslinks, M_c.

$$\xi = \left(lv_{2,s}^{-1/3}\right)\left(\frac{2C_n M_c}{M_r}\right)^{1/2} \quad (1.5)$$

Eqn (1.5) shows the distance between crosslinks, ξ, based on the swollen volume fraction of the polymer, $v_{2,s}$, the Flory characteristic ratio, C_n, the molecular weight between crosslinks, M_c, the average distance of one bond in the polymer backbone, l, and the molecular weight of polymer repeat unit, M_r.[17]

This physical length can be used to determine the diffusion coefficient of a solute through the hydrogel network. Comparing the molecular radius, a, to the distance between crosslinks in the hydrogel, ξ, one can calculate the amount by which the movement of the solute is slowed. Lustig and Peppas incorporated this principle into a solute diffusion model based on free volume theory:[27,62]

$$\frac{D_{gel}}{D_0} = \left(1 - \frac{a}{\xi}\right) e^{-\left(Y\left(\frac{v}{1-v}\right)\right)} \approx \left(1 - \frac{a}{\xi}\right) \qquad (1.6)$$

Eqn (1.6) shows the ratio of the diffusion coefficient within a hydrogel, D_{gel}, to the diffusion coefficient within free solution, D_0, based on molecular radius, a, and the average distance between crosslinks, ξ. Here, Y is the lumped free volume parameter, and v is the volume fraction of the polymer. This relation simplifies at low volume fractions since the exponential term becomes approximately 1.[27]

An example of diffusion from a source within synthetic hydrogels was described by Peret *et al.*[59] when protein-releasing microspheres were encapsulated within a small section of a hydrogel to create a localized source (Figure 1.2). In this work, the rest of the hydrogel acted as a sink for molecules released from the localized source.[59] The gradients that formed in these

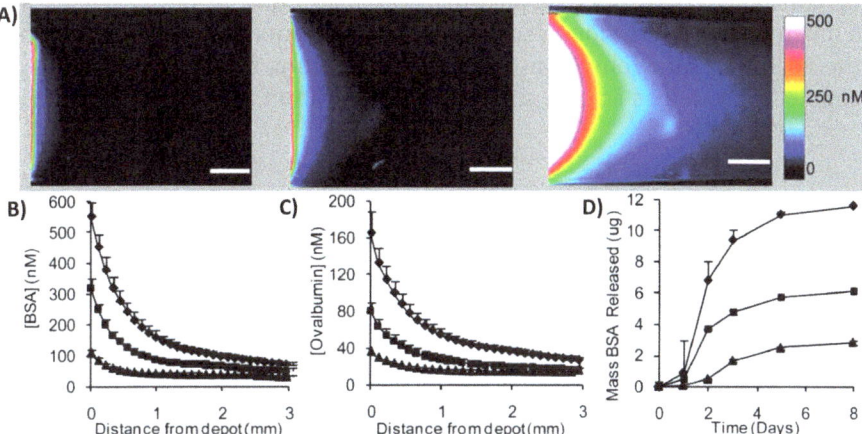

Figure 1.2 (A) Color-coded diagram of the concentration gradients within the hydrogel of bovine serum albumin (BSA). From left to right, the concentration of the BSA loading solution that the microspheres were incubated in increases (10, 30, and 60 mg mL^{-1}). (B) Quantified BSA concentration as a function of position (triangle is 10 mg mL^{-1}, square is 30 mg mL^{-1}, and diamond is 60 mg mL^{-1} loading solution). (C) Quantified ovalbumin concentration as a function of position. (D) Mass of BSA released from the gel construct over time. Reproduced with permission from ref. 59.

hydrogels were dependent upon source concentration and the hydrodynamic radii of the diffusing molecules.[59] Both of these dependencies can be predicted from equations described thus far in this chapter. Another study by Cruise et al.[60] demonstrated trends in the diffusion of proteins through hydrogel networks that were similar to those seen by Peret et al. In this work, the permeability of poly(ethylene glycol) (PEG) hydrogel networks to proteins of various sizes was controlled solely by adjustments to hydrogel network properties.[60]

1.3 Soluble Factor Generation and Consumption

1.3.1 Soluble Factor Generation and Consumption Within the Native ECM

Though the relationships discussed in previous sections can be used to generally explain gradient formation in the ECM, they make unrealistic assumptions about soluble factor generation and consumption. For instance, cells are only capable of producing soluble molecules at specific rates,[12,15,53] which Savinell et al. established to be on the order of thousands of molecules per cell per second.[63,64] Additionally, the number and distribution of cells producing a particular soluble factor is also a matter to consider, as well as any feedback mechanisms that may influence the production rate. The same statements can be made about soluble factor consumption. Although there are *in vivo* situations where soluble molecules are produced or consumed regardless of all other variables, this is rare.[65] Therefore, a much better description of soluble factor production and consumption are the following functions (eqn (1.7)), which account for cell number at the source, N, as well as soluble molecule concentration, C, and time, t:

$$C(x) = P'(N, C, t)$$
$$C(x) = -E'(N, C, t) \qquad (1.7)$$

Eqn (1.7) shows that the soluble factor concentration is dependent on the rate at which the soluble factor can be produced (top) or consumed (bottom) by the point-sources and sinks. P' and E' represent functions that relate soluble factor production and consumption to cell number, N; concentration of that factor, C; and time, t, to soluble factor concentration.

With the assumption that the rate of production is independent of concentration and time, this relationship is incorporated into eqn (1.1) as the y-intercept (eqn (1.8)). Of note is that if one assumes a time-dependence on this source, then eqn (1.1) is no longer valid as it was derived under a steady state assumption.

$$C(x) = P' - \frac{Jx}{D} \qquad (1.8)$$

Eqn (1.8) shows Fick's first law with a more realistic term, P', than C_0 for the soluble factor production term. Eqn (1.8) is fundamentally the same as eqn (1.1), except that now it includes a more realistic soluble factor production term. Variables such as the number of cells at the source can now be introduced. Since production is not necessarily constant over time as has been assumed here, the implications of temporal dynamics is discussed further in the following sections.

Besides including soluble factor production parameters, this equation can be further modified to include the fact that cells are not all condensed to a point. In many biologic examples, the source should not be approximated as a point and is actually distributed over a small area.[12,15,53] Including position dependence in the production term now changes the mathematical description to a new version of Fick's first law:

$$C_p = P'(N, C, t, x)$$

$$C_p \approx P'(x)$$

$$J = -D\frac{dC}{dx} + P'(x) \qquad (1.9)$$

$$C(x) = P(x) - \frac{Jx}{D}$$

Eqn (1.9) shows Fick's first law when the production term is dependent on position, x. Here $P(x)$ is equal to the integral of the $P'(x)$ function. The relationship between concentration and position is now defined in part by the production term, P. Depending on the nature of the relationship between soluble factor production and position, many different solutions are possible. Several hypothetical cases are shown in Figure 1.3a. The gradient is no longer necessarily linear as it is in all cases discussed thus far in the chapter.

Similar to the assumptions of a perfect point-source, the assumption of the perfect point-sink is not realistic, either. In fact, diffusion through a decellularized region to a perfect, point-sink is an unusual situation *in vivo* since it is understood that cellular consumption of soluble molecules is the principal method by which these molecules are cleared.[65] For instance, many soluble proteins bind to membrane-bound receptors, which are then internalized into the cells.[66,67] Other solutes, such as steroid hormones and other lipid-soluble molecules, are able to diffuse through the cell membrane and bind to intracellular components.[66,67] Cellular nutrients, such as O_2, are also consumed internally within the cells through various metabolic processes.[66,67] As a result, cells within the transport region should be included in the model, and it is assumed that these cells consume the soluble

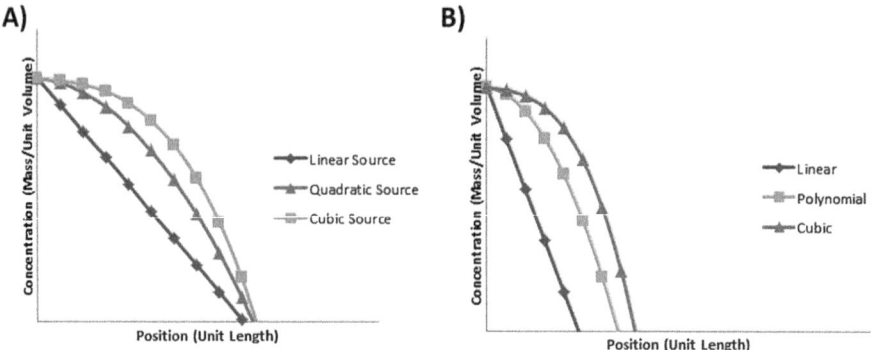

Figure 1.3 (A) Concentration gradients formed from production at a spatially distributed source. Possible source distributions shown are linear, quadratic, and cubic. Notably, if the source is distributed in a non-linear fashion then the formed gradient is also non-linear. (B) Concentration gradients produced by a spatially distributed source and consumed at a spatially distributed sink. While not meant to be biologically relevant, some example relations are shown here (linear indicates linear production and consumption functions, quadratic indicates quadratic relation for both, and cubic indicates a cubic relation for both). Notably the gradients span a much smaller distance than those shown in (A) because of the inclusion of an elimination term.

factor. Including this relationship into the original application of Fick's law gives:

$$C_e = E'(N, C, t, x)$$
$$C_e \approx E'(x)$$
$$\text{BC1}: C(0) = C_0 \qquad (1.10)$$
$$J = -D\frac{dC}{dx} + P'(x) - E'(x)$$
$$C(x) = (P(x) - E(x)) - \frac{Jx}{D}$$

Eqn (1.10) shows Fick's first law when the production and elimination terms are dependent on position, x. The concentration of soluble factors eliminated is represented by C_e. Here $P(x)$ is the integral of $P'(x)$ and $E(x)$ is the integral of $E'(x)$. Also, C_0 is lumped into the $P(x)$ function. This relationship shows a balancing act between the production, elimination, and diffusion terms. The dependence on position of the production and elimination terms, as well as the diffusion coefficient, determines the properties of the gradient (Figure 1.3b).

As discussed previously in this section, solutes are often consumed *via* interaction with cellular components.[66,67] Taking interaction kinetics into account can bring important improvements to the description of soluble factor consumption by cells.[65] We assume that the solute in question, referred to as the 'ligand', interacts with some type of cellular component, referred to as the 'receptor', and that this interaction is dependent on the concentration of both components. Many models have been derived that describe the kinetics of solutes binding to cellular components, and the general mathematical representation of such kinetics is:[68,69]

$$\frac{d(CR)}{dt} = k_{on}(R)(C) - k_{off}(CR) \quad (1.11)$$

Eqn (1.11) is a typical equation relating the number of receptor–ligand complexes, CR, over time, t, to the concentration of the receptor, R, soluble factor concentration, C, the binding constant, k_{on}, and the dissociation constant, k_{off}.[68,69]

Eqn (1.11) is derived for a soluble protein binding to extracellular receptors;[68,69] however, this general expression can be adapted for most situations of solutes binding to cellular components. Combining eqn (1.10) and (1.11) can be accomplished with the assumption of steady state and the assumption that position dependence is small relative to the concentration dependence. Another assumption made is that all terms (aside from the concentration terms) are not dependent on time or position.

$$E' = \frac{d(CR)}{dt} = k_{on}(R)(C) - k_{off}(CR) \approx 0$$

$$J = -D\frac{dC}{dx} + P' - k_{on}(R)(C) + k_{off}(CR) \quad (1.12)$$

$$C(x) = C_0 \, e^{\frac{-k_{on}(R)}{D}x} + \frac{k_{off}(CR) + P - J}{k_{on}(R)}\left(1 - e^{\frac{-k_{on}(R)x}{D}}\right)$$

Eqn (1.12) is the function for solute concentration, C, in terms of position, x, and it shows a dependence on solute concentration at the source, C_0, the binding and unbinding constants, k_{on} and k_{off}, the diffusion coefficient, D, the receptor concentration, R, and the concentration of receptor–ligand complexes, CR. Though incorporation of realistic source and sink terms can increase the complexity of the solution, these models remain in the form of linear differential equations and therefore can be solved without great difficulty. With the removal of perfect point-sink and source assumptions, a more realistic model of transport in the native ECM can be created. Additionally, the important assertion that these gradients are not necessarily linear is also made.

1.3.2 Soluble Factor Generation and Consumption Within Synthetic Hydrogels

Like sources and sinks within the native ECM, most sources and sinks within synthetic hydrogels are not accurately modelled as perfect, point entities, either. Additionally, since cells may serve as sources or sinks within synthetic hydrogels, the geometry of their placement can affect transport and gradient properties just as in the native ECM. Synthetic constructs such as microparticles, microfluidic channels, and biochemically functionalized regions of the hydrogel can also serve as important source or sink regions.

Since the geometry of the production and consumption regions can greatly affect soluble factor gradients (Section 1.2.1), there is a need to spatially pattern cellular populations within hydrogels. Technologies to pattern cells into complex geometries within a hydrogel have been utilized in several studies, and nearly all have utilized microfluidic or lithographic processes.[19,25,70] Other approaches involve encapsulating the cells into microgels, which can then be assembled into larger hydrogels.[19] More complex methods exist to pattern cells within synthetic hydrogels; however, some methods involve fabrication steps that have proven to be toxic to cells.[19]

Microfluidic constructs within hydrogels have also been utilized as synthetic sources and sinks. Basic applications of microfluidic channels within hydrogels have been used as nutrient sources and waste sinks, but more recently microfluidic channels have been used to model the *in vivo* vasculature.[19] The advantage of microfluidic approaches is that parameters such as soluble factor concentration and flow rate can be used to adjust gradient parameters within the hydrogel. The relationship between soluble factor concentration, flow rate, and soluble factor gradients is a well-known transport problem, involving both the Reynolds and Schmidt numbers.[54,71] Acting as a sink carrying soluble molecules away from the construct, a microfluidic channel behaves as follows:

$$J = \frac{3\pi RDC_0}{\Gamma\left[\frac{4}{3}\right]}\left(\frac{2}{9}\right)^{\frac{1}{3}} \mathrm{Re}^{\frac{1}{3}}\mathrm{Sc}^{\frac{1}{3}}\left(\frac{R}{L}\right)^{\frac{2}{3}}$$

$$\mathrm{Re} = \frac{2R\rho v}{\mu}$$

$$\mathrm{Sc} = \frac{\mu}{\rho D}$$

(1.13)

Eqn (1.13) describes the mass flux, J, through the walls of a tube of flowing fluid in terms of the Reynolds number (Re) and the Schmidt number (Sc). Here R is the radius of the tube, D is the diffusion coefficient of the solute, C_0 is the concentration of the solute in the fluid, L is the length of the tube, v is the velocity of the fluid, μ is the fluid viscosity, ρ is the fluid density, and Γ is the gamma function.[54,71] Equations similar to eqn (1.13) can be derived for most situations involving the flux of mass in or out of a tube of flowing fluid.[54,71] Eqn (1.13) demonstrates that as either the concentration of the solute at the tube boundary or the fluid flow velocity increases, the mass of solutes removed from the hydrogel increases.

Another method of creating soluble factor sources and sinks within hydrogels is the incorporation of microparticles. Microparticle incorporation has many advantages, including the ability to pattern them into complex arrangements.[29] Technologies exist that allow for the tailoring of molecule release from microparticles so that several different release rates and profiles are possible.[28] Some microparticle formulations are able to release molecules over extended periods of time (days to weeks) and maintain relatively constant release rates over this time frame.[28]

Functionalization of regions in the hydrogel network to deliver or sequester soluble molecules is yet another widely utilized approach. Significant work has been done on polymer functionalization, and many chemistries exist to bind biological molecules to synthetic polymers.[30] As the molecules are liberated from the polymer, they diffuse down their concentration gradients. Many of the release mechanisms employed are enzymatic or hydrolytic in nature,[30] but other methods have been established that utilize reversible binding of molecules to the hydrogel matrix.[31] These methods are discussed in detail in later sections.

1.4 Matrix Interactions

1.4.1 Interactions with ECM Components

Thus far in the chapter, all presented equations have assumed that diffusing solutes do not interact with the ECM components in any way other than 'sieving action'. This is inaccurate for many solutes, and ECM components routinely interact with diffusing solutes through charge interactions and molecular recognition.[14,72] Additionally, recent studies have implicated proteoglycans as crucial elements in the transport mechanisms of some soluble molecules due to charge interactions and sequestering.[73]

The simplest case of matrix–solute interactions is that of charge interactions (attraction and repulsion) between diffusing solutes and ECM components. Since most of the ECM is negatively charged because of the presence of proteoglycans,[11,16] molecules with a strong positive charge can have their movement retarded by their attraction to ECM components of opposite charge.[51] Molecules with a negative charge can also have their movement retarded by repulsive interactions.[51] Because the ECM can slow molecules due to sieving action and electrostatic interactions, Lieleg et al. described the ECM as an 'electrostatic bandpass'.[14]

Beyond electrostatic attraction and repulsion, many solutes exhibit binding regions that allow them to specifically bind to various ECM components.[72,74,75] Such mechanisms have been elucidated for solutes such as fibroblast growth factor (FGF), transforming growth factor beta (TFGβ), platelet-derived growth factor (PDGF), and vascular endothelial growth factor (VEGF).[72,74] Because of this binding, the ECM can then act as a reservoir of soluble molecules that can be released due to specific cues or to influence fluctuations in the concentration of soluble molecules.[72] Evidence also exists that sequestering of soluble molecules within the ECM may serve as an important method of regulation for soluble factor signalling and gradient formation.[72,74]

The contribution of the binding and release of soluble molecules from the ECM to mass transport models is an extension of the elimination and production functions that were defined earlier in the chapter (Section 1.2.1). The function defining the matrix binding and release of soluble molecules is dependent on the concentration of soluble molecules (C), the concentration of matrix binding sites (MR), and the distribution of the ECM components with respect to position (x). Modification of the equations proposed in Section 1.2.1 with an ECM binding and release term yields:

$$J = -D\frac{dC}{dx} + P'(C,x) - E'(C,x) + M'(C,MR,x) \quad (1.14)$$

Eqn (1.14) shows the addition of a matrix binding term to a model of soluble factor transport in the ECM. Here $M'(C, MR, x)$ is a function that relates the matrix binding of molecules from solution to soluble factor concentration, C, to the concentration of free binding sites in the matrix, MR, and position, x.

This reservoir of soluble molecules that accumulates in the ECM can be released by proteolytic activity degrading the ECM and thus freeing the bound molecules.[72,74] This can be accounted for by the use of a step function, u, which has a value of 1 when proteolytic elements are present and 0 when these elements are not present.

$$J = -D\frac{dC}{dx} + P'(C,x) - E'(C,x) + M'(C,MR,x) + uP'_{Matrix}(x) \quad (1.15)$$

Eqn (1.15) is the incorporation of the 'proteolytic liberation' term into eqn (1.14). This allows one to account for molecules freed from the matrix when proteolytic elements are present. Here u is the unit-step function and $P'_{Matrix}(x)$ is the production term of molecules liberated from the matrix as a function of position.

Thorne *et al.* provided an example of the effect of ECM binding on soluble factor transport (Figure 1.4). The diffusion of two proteins in the brain ECM was investigated: lactoferrin (Lf) and transferrin (Tf). Lf is known to bind to heparin sulfate proteoglycans whereas Tf does not.[75] Performing *in vivo* experiments, the authors observed retarded diffusion of Lf within the ECM compared to the diffusion of Tf (Figure 1.4).[75] From this, the importance of the effects of matrix interactions on diffusing solutes is clear, as well as the fact that such interactions must be accounted for to accurately model solute diffusion in the native ECM.

1.4.2 Interactions with Hydrogel Components

Solute interactions with the synthetic polymers used in hydrogels are different from interactions with the native ECM for many reasons. Obviously soluble biological molecules do not have binding domains for synthetic

Soluble Molecule Transport Within Synthetic Hydrogels 15

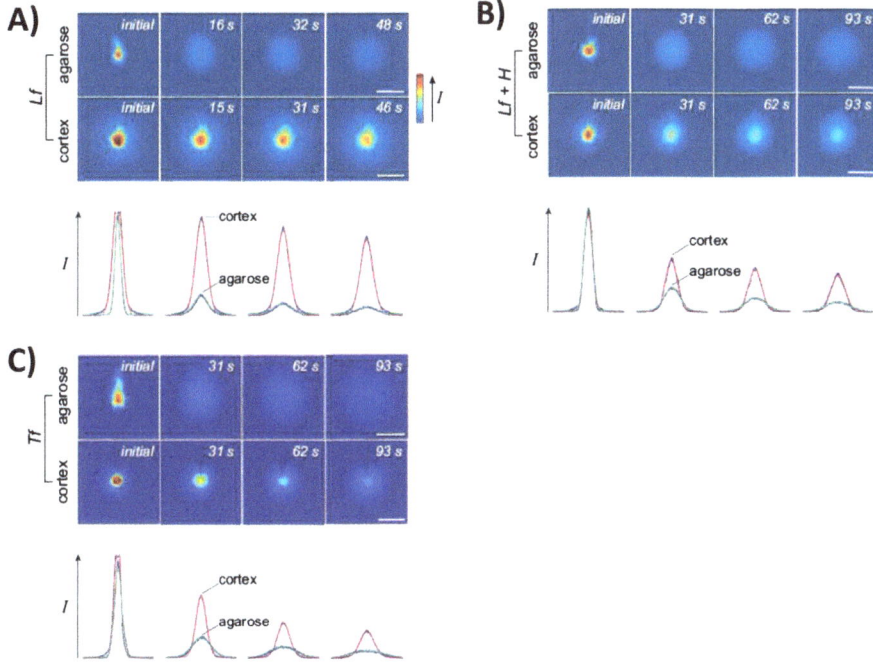

Figure 1.4 (A) Injection of Lf into the ECM of the brain cortex *versus* injection into an agarose substrate. Due to interactions between Lf and the ECM, the gradient is maintained over the course of the experiment. (B) Injection of Lf and solubilized H into the ECM of the brain cortex *versus* injection into an agarose substrate. As the Lf had already bound to the solubilized H, it diffused through the ECM quickly with few interactions. (C) Injection of Tf into the ECM of the brain cortex *versus* injection into an agarose substrate. As Tf does not interact with the ECM, it diffused away from the injection site quickly. H, heparin; Lf, lactoferrin; Tf, transferrin. In all parts, the *y*-axis is labeled with I for fluorescence intensity. Reproduced with permission from ref. 75.

polymers, and many synthetic polymers (such as poly(ethylene glycol)) typically have no specific interactions with diffusing solutes.[76] As a result, synthetic hydrogels do not recapitulate the specific binding and release mechanisms of the ECM.

However, interactions with ECM components can be incorporated into synthetic hydrogels through the use of functionalized biological molecules and peptides.[21,26,32–45,77–80] For example, in a study by Benoit *et al.* poly(ethylene glycol) hydrogels were modified with a chemically modified form of the glycosaminoglycan heparin (Figure 1.5a), an ECM component that has the ability to bind many soluble biological molecules.[21,26] The heparin-functionalized hydrogels caused increased osteogenic differentiation of mesenchymal stem cells relative to non-functionalized controls.[26] Other work by Benoit *et al.* showed that heparin-functionalized hydrogels were

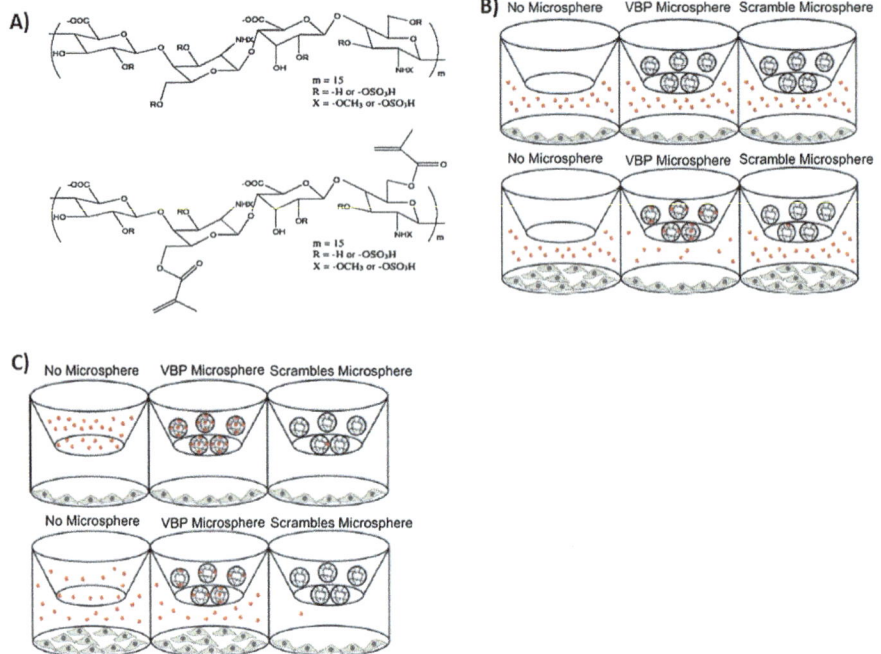

Figure 1.5 (A) Example method of heparin incorporation into a synthetic hydrogel. Here, heparin is modified with methacrylate groups so it can be incorporated into a hydrogel (heparin shown at top, modified heparin shown at bottom). Reproduced with permission from ref. 21. (B) Diagram of VEGF-binding hydrogel microparticles specifically binding VEGF from cell culture media and lowering the level of cell proliferation. Reproduced with permission from ref. 31. (C) Diagram of VEGF-binding hydrogel microparticles releasing VEGF into cell culture media and inducing cell proliferation. Reproduced with permission from ref. 31.

capable of binding soluble FGF and releasing it over a 5 week time frame.[21] Additionally, other studies have introduced the concept of using a peptide to bind heparin to biomaterials rather than incorporating it directly.[32–34]

Though a molecule like heparin can bind large quantities of soluble biological molecules, it cannot bind only specific biological molecules of interest. To accomplish the goal of binding specific molecules, work has been done with binding peptides that have been modelled after the heparin binding[35] or phage-display against specific molecules.[36] Though the binding capabilities of these peptides are well established, the extent of specificity of these peptides is less clear. An intriguing approach is the design of peptides based on cellular receptors,[37] as receptor interactions with soluble biological molecules are typically high affinity and highly specific. Such peptides were recently utilized in poly(ethylene glycol) hydrogel microspheres that were shown to bind and release VEGF, as well as inducing a biologic effect by sequestering or releasing VEGF during *in vitro* cell culture (Figure 1.5b and c).[31]

Soluble Molecule Transport Within Synthetic Hydrogels 17

While the use of both peptides and functionalized ECM components shows promise, how these components affect transport and gradient formation within the hydrogel network remains to be fully investigated. For instance, few studies examine the effects of binding peptides on gradient formation between a well-defined source and sink within a synthetic hydrogel.[72,74] Additionally, though the peptide and glycosaminoglycan components of the synthetic hydrogels discussed here are susceptible to proteolytic degradation, the effect of degradation on soluble factor binding and release have yet to be characterized.

1.5 Temporal Dependencies

1.5.1 Temporal Dependencies During *In Vivo* Transport

Thus far in this chapter, all equations have been derived at steady state. While such an approach is not invalid for all *in vivo* scenarios,[12,15,53,81] most *in vivo* situations involve time dependency. For instance, the reaction-diffusion models that have been used to describe many situations in developmental biology rely on chemical reactions between two or more species,[81] which occur at specific rates that are not necessarily constant with time. Another case where steady-state assumptions do not hold is the establishment of a soluble gradient from one group of cells to another. The cells functioning as the source cannot suddenly put forth a gradient of soluble molecules; it must be established over time as the soluble factor is produced.[12,15,53,63,64,81]

To begin the discussion, we examine Fick's second law of diffusion. This law states the concentration, C, changes not only with position, x, but also with respect to time, t:

$$\frac{dC}{dt} = -D\frac{d^2C}{dx^2}$$
$$0 = -D\frac{d^2C}{dx^2} \text{ at steady state} \quad (1.16)$$

Eqn (1.16) is Fick's second law of diffusion, where C is concentration, D is the diffusion coefficient, t is time, and x is position.[71] Fick's second law returns to Fick's first law when steady state is assumed. Returning to eqn (1.7), it is reasonable to assert that both the production, P'', and consumption, E'', terms are dependent on time:

$$\frac{dC}{dt} = -D\frac{d^2C}{dx^2} + P''(C,x,t) - E''(C,x,t) \quad (1.17)$$

Eqn (1.17) is (1.7) reformulated with a dependence on time assumed on both the production, P'', and elimination, E'', terms. Eqn (1.17) is more complex than all previously proposed equations because the production and elimination terms show a dependence on position, x, as well as time, t.

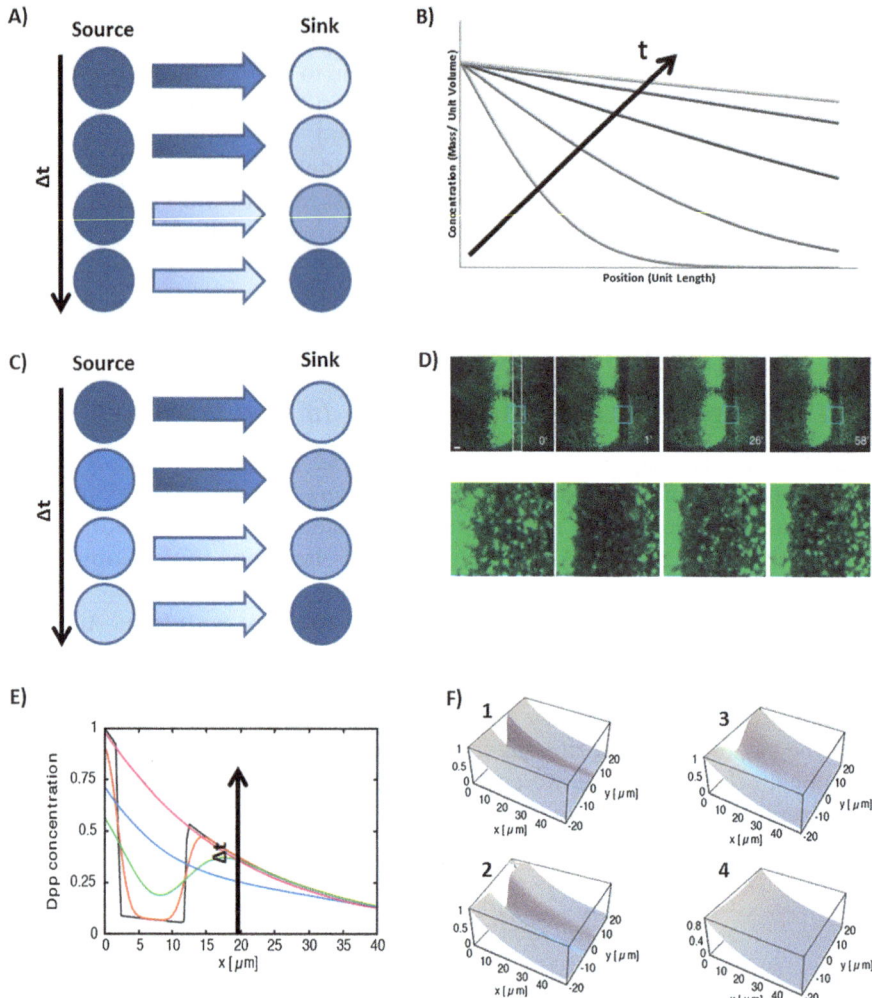

Figure 1.6 (A) Diagram of the proposed time-dependent problem. Here, the sink has a finite ability to sequester the solute but the source has an infinite amount of solute. (B) Concentration gradient changing over time due to the saturation of the sink region. (C) A more realistic diagram of the problem where not only does the sink become saturated but the source becomes depleted, as well. (D) Fluorescent data gathered in *Drosophila* from FRAP experiments on the soluble factor Dpp. The region of interest for the experiment is indicated by the white box in the upper left-most image. The numbers in the lower right-hand corner of the images indicates the time (in minutes) after the experiment has begun, with the left-most image being taken immediately before the experiment began. Second row of images is a close-up of the region indicated by the blue box in the first row. Scale-bar in the upper row of images represents a distance of 10 μm. (E) Theoretical curves of gradient formation in *Drosophila* based on gathered FRAP data. Progression of time through the curves is indicated on the graph by the black arrow. (F) 3D

The focus of this chapter is not on solving partial differential equations, so the reader should consult other references for the solutions to such problems.[82] To simplify eqn (1.17), the production terms are neglected as well as any dependence on time or position for the elimination function (Figure 1.6a). This leaves the following well-studied problem of reaction and diffusion from a constant source:[65]

$$\frac{dC}{dt} = -D\frac{d^2C}{dx^2} - kC$$

BC 1 : $C(x, 0) = 0$ for $x > 0$ and $t = 0$

BC 2 : $C(0, t) = C_0$ for $x = 0$ and $t > 0$

BC 3 : $C(\infty, t) = 0$ for $x \to \infty$ and $t > 0$ (1.18)

$$\frac{C}{C_0} = \frac{1}{2}\exp\{-x\sqrt{k/D}\}\mathrm{erfc}\left\{\frac{x}{2\sqrt{Dt}} - \sqrt{kt}\right\}$$
$$+ \frac{1}{2}\exp\{x\sqrt{k/D}\}\mathrm{erfc}\left\{\frac{x}{2\sqrt{Dt}} + \sqrt{kt}\right\}$$

Eqn (1.18) shows reaction and diffusion from a constant source. Here C is concentration, x is position, t is time, D is the diffusion coefficient, and k is the rate constant of the elimination reaction; erfc denotes the complementary error function and exp is the exponential function.[65] Eqn (1.18) is represented graphically in Figure 1.6b. A more realistic version of this problem would not only have a sink whose concentration is increasing over time but a source whose concentration falls over time. Such a situation is diagrammed in Figure 1.6c. The modelling of this situation is complex, so it is not discussed here further; however, it is important to understand that this type of complexity can occur in both natural and synthetic systems when temporal dynamics are taken into account.

Experiments using fluorescence recovery after photobleaching (FRAP) techniques offer insight into the temporal dynamics of gradient formation *in vivo*.[13] Using FRAP in conjunction with fluorescently tagged proteins, experimenters can track protein transport and gradient formation within an organism. Kicheva *et al.* used FRAP to evaluate the kinetics of decapentaplegic (Dpp) gradient formation in *Drosophila* (Figure 1.6d–f).[13] Their experimental approach involved the photobleaching of a chosen area next to

---

representation of theoretical curves shown in (E). Images are numbered to show the progression of time through the experiment. Image 1 indicates the time point immediately after photobleaching, and Image 4 shows full recovery of the gradient. Reproduced with permission from ref. 13.

the source region and observing the return of fluorescence to the region.[13] Based on data obtained from the FRAP technique, they were able to estimate the various parameters involved in Dpp gradient formation: the diffusion constant, the soluble factor degradation rate, and production rate at the source.[13]

Most models of transport in the native ECM do take into account temporal dynamics since nearly all *in vivo* processes are time dependent.[81] For instance, the rate of soluble factor production and consumption, unless perfectly matched, cause a system to move away from steady state. Other complications include changes in the number of cells that are consuming, producing, or otherwise altering soluble factor concentration; enzymatic effects on the soluble molecules themselves, on cellular receptors, or on ECM molecules; and secretion of additional ECM by cells. While incorporation of all of these aspects into a mathematical model may greatly increase its complexity, the overall point is clear that the most realistic representation of *in vivo* transport includes temporal dynamics.

1.5.2 Temporal Dependencies During Transport in Synthetic Hydrogels

As for steady-state assumptions in the native ECM, assumptions of steady state are not entirely invalid for synthetic hydrogels; however, the most realistic mathematical models of soluble factor diffusion in synthetic hydrogels take temporal dynamics into account. While many synthetic hydrogels have been shown to be stable when chemically crosslinked,[38–41,83] the hydrogel environment is not necessarily static. Cells encapsulated in the hydrogel can secrete their own ECM, proliferate, and alter their rates of soluble factor production and consumption. Furthermore, hydrogel components can be engineered to change over time *via* hydrolytic, enzymatic, or external mechanisms (such as pH or temperature).

A widely used method of altering synthetic hydrogels over time involves hydrolytic degradation. Many chemistries have been employed,[30,42–44] and many of these allow the experimenter to modulate the rate of degradation. The space freed by the hydrolytic degradation can lead to an increase in cell spreading and viability within a hydrogel.[30,38,39,42–44] Hydrolytic degradation has also been linked to increased cellular production of ECM proteins within the hydrogel.[30,38,39,42–44]

Besides altering cellular viability and ECM production, hydrolytic degradation also affects soluble molecule transport within the hydrogel. Returning to eqn (1.6), the permeability of the hydrogel to a solute is dependent on the average distance between crosslinks, ξ. As the hydrogel degrades, ξ increases. Additionally, a degrading hydrogel changes its concentration (*i.e.* volume fraction) of hydrogel components over time as well as changes in the system parameters. Alterations to the distance between crosslinks alters the diffusion coefficients of solutes within the hydrogel (eqn (1.6) and (1.19)), and

Soluble Molecule Transport Within Synthetic Hydrogels

thus provides an opportunity to modulate soluble cell–cell signalling and gradient formation within these hydrogels.

$$\frac{D_{\text{gel}}}{D_0} = \left(1 - \frac{a}{\xi(t)}\right) e^{-\left(Y(t)\left(\frac{v(t)}{1-v(t)}\right)\right)} \quad (1.19)$$

Eqn (1.19) is the restated Lustig–Peppas model of the diffusion constant within a hydrogel where the hydrogel properties are functions of time. Eqn (1.19) is complex since several variables are related to time, such as the distance between crosslinks, ξ, the polymer volume fraction, v, and the free volume theory diffusion parameters, Y. However, work by Anseth *et al.* on hydrolytically degrading PEG-based hydrogels used an approximation that is valid for highly swollen gels (*i.e.* the volume fraction can be approximated as zero):[39]

$$\text{Given: } \frac{a}{\xi(t)} \approx (M_c(t))^{-\frac{7}{10}} \text{ and } Q(t) \approx (M_c(t))^{3/5}$$

$$\frac{D_{\text{gel}}}{D_0} = \left(1 - \frac{a}{\xi(t)}\right) e^{-\left(Y(t)\left(\frac{v(t)}{1-v(t)}\right)\right)} \approx \left(1 - \frac{a}{\xi(t)}\right) e^{-(0)}$$

$$1 - \frac{D_{\text{gel}}}{D_0} = \frac{a}{\xi(t)} \approx (M_c(t))^{-7/10} \quad (1.20)$$

$$1 - \frac{D_{\text{gel}}}{D_0} \approx (Q(t))^{-7/6}$$

Eqn (1.20) is the simplified relation of the diffusion constant in highly swollen hydrogels to hydrogel parameters. Here Q is the mass swelling ratio of the hydrogel (wet mass over dry mass).[39] Eqn (1.20) relates the changing diffusion coefficient of a solute in the hydrogel to the molecular weight between crosslinks, M_c, and ultimately to the mass equilibrium swelling ratio of the hydrogel, Q.[39] A comparison of this model's predictions to experimental data on the release of proteins from the hydrogel was done to validate this model.[39] A potential limitation of this approach, however, is that the radius of the solute is neglected. Furthermore, this model is only valid for highly swollen hydrogels; if this is not the case, then the modelling of these relations returns to eqn (1.19).[39]

Hydrolytic degradation as a means to modulate soluble factor transport within hydrogels is not without drawbacks. The most obvious is the difficulty in matching hydrolytic degradation rates to the timescale of biologic events,[45] such as cellular migration within a hydrogel.[46] As a result, significant research has been done to develop hydrogels that degrade *via* cell-mediated mechanisms. Many cell-mediated degradation strategies involve the cross-linking of the hydrogel with molecules that are susceptible to enzymatic cleavage.[30,39–44,83] These studies demonstrate cell viability, proliferation, and ECM production by cells within these hydrogels.[40,41]

Two important differences between enzymatic and hydrolytic degradation mechanisms are:

- Degradation is dependent on time and on enzyme concentration.[39,83]
- The kinetics of the enzymatic degradation are dependent on several variables including concentration of the degradable unit, half-life of the enzyme in the active form, cellular production of the enzyme, and the rate at which the degradable unit can be cleaved by the enzyme.[39,83,84]

Rice et al. derived a model to describe the degradation of a synthetic hydrogel in the presence of an exogenously delivered enzyme.[83] During the course of this degradation (concentration of degradable crosslinker is represented by the variable C), the enzyme (whose concentration is represented by the variable E) is involved in the following reactions:

$$E + C \overset{k_1}{\leftrightarrow} E^*C \overset{k_2}{\to} E + D$$
$$E \overset{k_d}{\to} E_{\text{inactive}} \tag{1.21}$$

Eqn (1.21) shows the chemical reactions involved in the enzymatic degradation of a synthetic hydrogel. Here E is enzyme concentration, C is the degradable crosslinker concentration, D is the concentration of degradation products, E_{inactive} is the concentration of inactivated enzyme, and E^*C is the enzyme–crosslinker complex. The rate constants k_1, k_2, and k_d are for reaction 1, reaction 2, and the enzyme inactivation reaction respectively.[83]

The following differential equations can be solved to yield an expression for the ratio of intact crosslinker molecules at a given time to the original number of crosslinker molecules, N/N_0:

$$\frac{dE}{dt} = -(k_1)(C)(E) + (k_{1r})(E^*C) + (k_2)(E^*C) - (k_d)(E)$$

$$\frac{dC}{dt} = -(k_1)(C)(E) + (k_{1r})(E^*C) \tag{1.22}$$

$$\frac{N}{N_0} = e^{\frac{(k^*)(E_0)}{(k_d)}\left(e^{-(k_d)t} - 1\right)}$$

From eqn (1.21), the relations given in eqn (1.22) can be constructed and solved for the ratio of intact crosslinker at a given time to the original amount of crosslinker, N/N_0. Note that rate constants denoted with an 'r' indicate that they are the reverse rate constant and k^* indicates lumped enzyme kinetic parameters.[83]

Though not done by the authors of this study, the amount of intact crosslinker may be related to solute diffusion coefficients through eqn (1.20). Notably, this derivation assumes that the degradation reaction, and not the diffusion of the enzyme, is the limiting factor in the rate of hydrogel degradation.[83]

While there are numerous studies that discuss hydrogel degradation *via* hydrolytic or cell-mediated mechanisms for controlled release applications, there is a relative lack of studies using these technologies to study cellular signalling within the hydrogel environment. Additionally, for studies that do incorporate cells into hydrolytically and enzymatically degrading constructs, it is routinely noted that the cells secrete their own matrix.[30,38–44] How this secreted matrix affects the transport properties within the hydrogel has yet to be investigated.

Aside from degradation *via* hydrolytic and enzymatic mechanisms, work has also been done to develop 'responsive hydrogels', which allow for reversible changes to occur in the hydrogel based upon changes in the polymers that comprise the hydrogel. Such changes ultimately affect the distance between crosslinks, ξ, and thus affect the transport of solutes within the hydrogel. These strategies revolve around changes to solution pH, solution ionic strength, temperature, or the presence of a specific chemical compound.[30] All of these approaches combine the possibilities of experimenter-controlled changes, cellular-mediated changes, and reversibility of these changes.

The most ubiquitous of these designs are pH- and ionic strength-responsive hydrogels.[30] These hydrogels can be described on the basis of changes to the Gibbs free energy:[30]

$$G_{\text{Total}} = G_{\text{Mixing}} + G_{\text{Elastic}} + G_{\text{Osmotic}} + \ldots \tag{1.23}$$

Eqn (1.23) shows the Gibbs free energy of swelling for hydrogels in terms of the free energies from mixing, elastic retraction of the polymer chains, osmotic pressures, *etc.*[17,30]

Several models can predict the response of hydrogels designed to respond to pH or ionic strength changes.[30] A model derived by Peppas and colleagues relates the changes to diffusion coefficients within a hydrogel swelling in response to pH or ionic strength changes:

$$D_{\text{gel}} = D_0 e^{(\nu\alpha)} \tag{1.24}$$

Eqn (1.24) is the relation of diffusion coefficient, D_{gel}, in a pH or ionic dependent hydrogel based on the initial diffusion constant (D_0), the polymer volume fraction (ν), and the water–polymer interaction parameter (α).[30]

For hydrogels that change with respect to temperature (and not pH or ionic strength), an alternative expression has been developed:

$$\frac{D_{\text{gel}}}{D_0} = \frac{(1-\nu)^3}{(1+\nu)^2} \tag{1.25}$$

Eqn (1.25) shows the diffusion coefficient of a solute in a hydrogel, D_{gel}, swelling in response to temperature changes. Here D_0 is the original diffusion coefficient of the solute in the gel and ν is the polymer volume fraction.[30] Since volume fraction of polymer within the hydrogel, ν, is easily measured as a function of temperature, this is a convenient relation.[30] Though designs do exist that combine the ideas of pH, ionic strength, and temperature changes, mathematical representations of them remain to be developed.[30] Furthermore development of hydrogels that respond to specific chemical compounds remains particularly challenging, and although some recent approaches have shown limited 'bioresponsiveness', there are not yet generic mechanisms for response to a chemical compound of interest. Indeed, controlled bioresponsiveness represents a grand challenge in design of synthetic hydrogels.

1.6 Convection

1.6.1 *In Vivo* Transport by Convection

Convective flow is an aspect of *in vivo* transport that is neglected in many models. This is because convection is considered to be only a small component of transport in the ECM due to the slow flow velocities.[85] For instance, in the lymphatic system, flow velocities of 0.1 to 1.0 µm s^{-1} have been observed.[85] However, evidence exists that these small interstitial flows serve to make important alterations to gradients *in vivo*, such as in developing embryos, lymphangiogenesis, and capillary morphogenesis.[85]

To incorporate convective flow into transport equations, Fick's second law (eqn (1.16)) is modified as follows:[71]

$$\frac{dC}{dt} + V\frac{dC}{dx} = -D\frac{d^2C}{dx^2} + P'' + E'' \tag{1.26}$$

Eqn (1.26) is a modified transport equation to incorporate mass transport due to convective flows.[71] In eqn (1.26), a new parameter has been added, the flow velocity, V. As a result, flows in the ECM must be known in order for eqn (1.26) to be utilized. Flow in the ECM can be modelled *via* the Brinkman equation for flow through porous media:[85,86]

$$P_x = -\frac{\mu}{K}V + \mu V_{xx} \tag{1.27}$$

Eqn (1.27) is the Brinkman equation for flow within a porous media. Here μ is the fluid viscosity, K is the media permeability, P_x is the derivative of the pressure function with respect to x, and V_{xx} is the second derivative of the velocity distribution with respect to x.[86]

Eqn (1.27) is a well-studied relationship and many solutions are available in the literature. A study by Fleury *et al.* utilized known solutions to eqn (1.27) and found that even slow interstitial flow rates (0.1–6.0 µm s^{-1}) impart significant bias to a soluble gradient in the ECM (Figure 1.7).[85] Their model included ECM binding of, cell production and consumption of, and protease-mediated liberation of molecules from the ECM.[85] They noted that high levels of ECM sequestering of solutes did not overcome this effect, as both the soluble molecule diffusing to matrix binding sites and the protease to cleave it from such sites were biased by the convective flow.[85] Such studies

Soluble Molecule Transport Within Synthetic Hydrogels 25

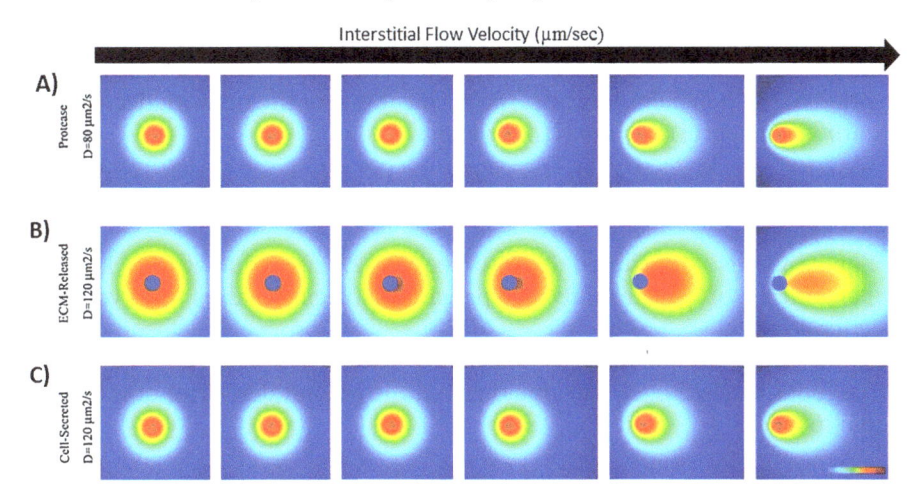

Figure 1.7 Simulated effects of interstitial convective flows on gradient formation *in vivo*. Image is color-coded with blue being the lowest concentration and red being the highest. Note that the diffusion coefficient assumed for the species in each row is indicated at left. (A) Proteases capable of liberating soluble molecules from the ECM. (B) Soluble molecules released from the ECM by these proteases. (C) Soluble molecules secreted by cells are also diffusing through the matrix in a biased fashion due to interstitial flows. Reproduced with permission from ref. 85.

demonstrate that while convective flows may not be the principle component creating soluble factor gradients in the native ECM, they may alter the shape of these gradients.

1.6.2 Transport in Synthetic Hydrogels by Convection

As in the native ECM, convection within synthetic hydrogels is assumed to be negligible except in special circumstances, such as the presence of macro-scale pores within the hydrogel or forced flow through the hydrogel.[70] However, to determine whether or not the convection is significant in any situation, eqn (1.27) can be used to assess the flow velocity based upon the permeability of the hydrogel, K. Many models have been derived to determine K in polymeric materials,[87] and for the purpose of this section, we use the general model proposed by Jackson and James.[87,88] However, the reader should note that more specific models exist.

$$K = r^2 \frac{3}{20v}(-\ln(v) - 0.931) \qquad (1.28)$$

Eqn (1.28) is the permeability within a synthetic hydrogel matrix based on v, the volume fraction of the polymer, and r, the average radius of the polymer molecules that compose the matrix.[87] Eqn (1.28) shows a relation between the polymer volume fraction, v; the radius of the polymer molecules that compose the matrix, r; and the network permeability, K. As the volume fraction of the polymer increases, the permeability of the network decreases. Incorporating eqn (1.28) into (1.27) yields:

$$P_x = -\frac{\mu}{K(r,v)} V + \mu V_{xx} \qquad (1.29)$$

Eqn (1.29) is a modified form of eqn (1.27) introducing the known dependencies of permeability on polymer molecule radius and volume fraction.

A further simplification is that the velocity is relatively constant throughout the hydrogel, thus making V_{xx}, the second derivative of velocity, equal to zero:

$$P_x = -\frac{\mu}{K(r,v)} V \qquad (1.30)$$

Eqn (1.30) is a simplified form of eqn (1.29) when the velocity is nearly constant. Eqn (1.30) shows that decreasing permeability causes the pressure needed to maintain a given velocity to increase. Conversely, increasing permeability causes a decrease in the pressure needed to maintain a given velocity.

Kapur *et al.* have measured the permeability of polyacrylamide hydrogels.[89] The authors of this study forced water through the hydrogel with pressurized nitrogen and measured the velocity of the water as it moved through the hydrogel.[89] Permeability was dependent on the polymer volume fraction,[89] which is predicted by eqn (1.28). An alternative technique, proposed by Lin *et al.*, involved estimating hydrogel permeability based upon mechanical properties of the hydrogel network.[90] Despite this work, however, convective transport in hydrogels is typically ignored, and its effects on transport within the hydrogel remain to be fully investigated.

1.7 Future Directions and Concluding Remarks

Soluble transport within the native ECM is complex, involving multiple populations of cells, matrix interactions, temporal dependencies, and convective flows. These complexities make the study of transport mechanisms *in vivo* a difficult task. Therefore, there is a great demand for *in vitro* methods to study these transport mechanisms. To address this demand, many different synthetic hydrogel technologies have been developed that aim to recapitulate aspects of *in vivo* soluble transport.

Future work will likely involve more intricate characterization of transport within these hydrogel environments, since many of these technologies have only been investigated from a standpoint of delivering solutes to the environment outside of the hydrogel. For instance, hydrogels that are engineered to degrade over time have been extensively explored from a drug delivery standpoint, but few studies have examined how the increasing distance between crosslinks allows for changes in soluble transport and gradient formation within the hydrogel. Another example is the use of hydrogels containing peptides that can bind specific soluble molecules. Although these hydrogels can deliver solutes to locations outside the hydrogel, little attention has been given to how these peptides affect soluble transport and gradient formation within the hydrogel.

Beyond characterizing the direct effects that many of these technologies have on transport within synthetic hydrogels, some fundamental aspects remain to be fully studied. For instance, although hydrogels have been designed respond to multiple types of stimuli at once, mathematical models to describe and ultimately predict their behaviour do not yet exist.[30] Convective flows within hydrogels are also typically ignored, and their effect on soluble transport within hydrogels has yet to be fully characterized.[70] Finally, patterning of source and sink regions (*e.g.* cell populations) within hydrogels into complex arrangements remains difficult, as many synthetic techniques are toxic to cells.[19] Therefore, although much progress has been made in the development of synthetic hydrogels to recapitulate aspects of native ECM soluble transport, much research remains to be done.

References

1. T. L. Chen, W. Shen and F. B. Kraemer, *J. Cell. Biochem.*, 2001, **82**, 187–199.
2. C. M. Shea, C. M. Edgar, T. A. Einhorn and L. C. Gerstenfeld, *J. Cell. Biochem.*, 2003, **90**, 1112–1127.
3. J. D. Abbott, Y. Huang, D. Liu, R. Hickey, D. S. Krause and F. J. Giordano, *Circulation,* 2004, 3300–3305.
4. T. Ara, Y. Nakamura, T. Egama, T. Sugiyama, K. Abe, T. Kishimoto, Y. Matsui and T. Nagasawa, *Proc. Natl. Acad. Sci. U. S. A.*, 2003, 5319–5323.
5. Y. Jung, J. Wang, A. Schneider, Y. X. Sun, A. J. Koh-Paige, N. I. Osman, L. K. McCauley and R. S. Taichman, *Bone*, 2006, **38**, 497–508.
6. C. Urbich, A. Aicher, C. Heeschen, E. Dernbach, W. K. Hofmann, A. M. Zeiher and S. Dimmeler, *J. Mol. Cell. Cardiol.*, 2005, **39**, 733–742.
7. T. Tabata and Y. Takei, *Development*, 2004, **131**, 703–712.
8. D. M. Duprez, K. Kostakopoulou, P. H. Francis-West, C. Tickle and P. M. Bricknell, *Development,* 1996, **122**, 1821–1828.
9. P. H. Francis, M. K. Richardson, P. M. Bricknell and C. Tickle, *Development*, 1994, **120**, 209–218.
10. C. J. Dowd, C. L. Cooney and M. A. Nugent, *J. Biol. Chem.*, 1998, **274**, 5236–5244.
11. L. Kjellen and U. Lindahl, *Annu. Rev. Biochem.*, 1991, **60**, 443–475.
12. A. D. Lander, Q. Nie and F. Y. M. Wan, *J. Comput. Appl. Math.*, 2006, **190**, 232–251.
13. A. Kicheva, P. Pantazis, T. Bollenbach, Y. Kalaidzidis, T. Bittig, F. Julicher and M. Gonzalez-Gaitan, *Science,* 2007, **315**, 521–525.

14. O. Lieleg, R. M. Baumga and A. R. Bausch, *Biophys. J.*, 2009, **97**, 1569–1577.
15. K. Page, P. K. Maini and N. A. M. Monk, *Physica D*, 2002, **181**, 80–101.
16. E. Ruoslahti, *Annu. Rev. Cell Biol.*, 1988, **4**, 229–255.
17. N. A. Peppas, J. Z. Hilt, A. Khademhosseini and R. Langer, *Adv. Mater.*, 2006, **18**, 1345–1360.
18. K. S. Anseth, C. N. Bowman and L. Brannon-Peppas, *Biomaterials*, 1995, **17**, 1647–1657.
19. H. Geckil, F. Xu, X. Zhang, S. Moon and U. Demirci, *Nanomedicine*, 2010, **5**, 469–484.
20. C. R. Nuttelman, M. C. Tripodi and K. S. Anseth, *Matrix Biol.*, 2005, **24**, 208–218.
21. D. S. W. Benoit and K. S. Anseth, *Acta Biomater.*, 2005, **1**, 461–470.
22. S. J. Bryant and K. S. Anseth, *J. Biomed. Mater. Res.*, 2001, **59**, 63–72.
23. K. W. Kavalkovich, R. E. Boynton, J. M. Murphy and F. Barry, *In Vitro Cell. Dev. Biol.*, 2002, **38**, 457–466.
24. C. R. Nuttelman, M. C. Tripodi and K. S. Anseth, *J. Biomed. Mater. Res., Part A*, 2003, **68**, 773–782.
25. M. S. Hahn, J. S. Miller and J. L. West, *Adv. Mater.*, 2006, **18**, 2679–2684.
26. D. S. W. Benoit, A. R. Durney and K. S. Anseth, *Biomaterials*, 2007, **28**, 66–77.
27. B. Amsden, *Macromolecules*, 1998, **31**, 8382–8395.
28. M. Biondi, F. Ungaro, F. Quaglia and P. A. Netti, *Adv. Drug Delivery Rev.*, 2008, **60**, 229–242.
29. S. Sant, M. J. Hancock, J. P. Donnelly, D. Iyer and A. Khademhosseini, *Can. J. Chem. Eng.*, 2010, **88**, 899–911.
30. C. Lin and A. T. Metters, *Adv. Drug Delivery Rev.*, 2006, **58**, 1379–1408.
31. N. A. Impellitteri, M. W. Toepke, S. K. L. Levengood and W. L. Murphy, *Biomaterials*, 2012, **33**, 3475–3484.
32. G. A. Hudalla, J. T. Koepsel and W. L. Murphy, *Adv. Mater.*, 2011, **23**, 5415–5418.
33. G. A. Hudalla, N. A. Kouris, J. T. Koepsel, B. M. Ogle and W. L. Murphy, *Integr. Biol.*, 2011, **3**, 832–842.
34. C. Lin and K. S. Anseth, *Adv. Funct. Mater.*, 2009, **19**, 2325–2331.
35. H. D. Maynard and J. A. Hubbell, *Acta Biomater.*, 2005, **1**, 451–459.
36. S. M. Willerth, P. J. Johnson, D. J. Maxwell, S. R. Parsons, M. E. Doukas and S. E. Sakiymam-Elbert, *J. Biomed. Mater. Res., Part A*, 2006, **80**, 13–23.
37. C. Piossek, K. H. Thierauch, J. Schneider-Mergener, R. Volkmer-Engert, M. F. Bachmann, T. Korff, H. G. Augustin and L. Germeroth, *Thromb. Haemostasis*, 2003, **90**, 501–510.
38. G. A. Hudalla, T. S. Eng and W. L. Murphy, *Biomacromolecules*, 2008, **9**, 842–849.
39. K. S. Anseth, A. T. Metters, S. J. Bryant, P. J. Martens, J. H. Elisseeff and C. N. Bowman, *J. Controlled Release*, 2002, **78**, 199–209.
40. B. D. Fairbanks, M. P. Schwartz, A. E. Halevi, C. R. Nuttelman, C. N. Bowman and K. S. Anseth, *Adv. Mater.*, 2009, **21**, 5005–5010.

41. B. K. Mann, A. S. Gobin, A. T. Tsai, R. H. Schmedlen and J. L. West, *Biomaterials,* 2001, **22**, 3045–3051.
42. J. L. Drury and D. J. Mooney, *Biomaterials,* 2003, **24**, 4337–4351.
43. G. D. Nicodemus and S. J. Bryant, *Tissue Eng., Part B,* 2008, **14**, 149–165.
44. C. R. Nuttelman, M. A. Rice, A. E. Rydholm, C. N. Salinas, D. N. Shah and K. S. Anseth, *Prog. Polym. Sci.,* 2008, **33**, 167–179.
45. J. Zhu, *Biomaterials,* 2010, **31**, 4639–4656.
46. M. P. Lutolf, J. L. Lauer-Fields, H. G. Schmoekel, A. T. Metters, F. E. Weber, G. B. Fields and J. A. Hubbell, *Proc. Natl. Acad. Sci. U. S. A.,* 2003, **100**, 5413–5418.
47. W. J. King, M. W. Toepke and W. L. Murphy, *Acta Biomater.,* 2010, **7**, 975–985.
48. R. J. Phillips, W. M. Deen and J. F. Brady, *AIChE J.,* 1989, **35**, 1761–1769.
49. A. G. Ogston, B. N. Preston and J. D. Wells, *Proc. R. Soc. London,* 1973, **333**, 297–316.
50. F. Yuan, A. Krol and S. Tong, *Ann. Biomed. Eng.,* 2001, **29**, 1150–1158.
51. T. Stylianopoulos, M. Poh, N. Insin, M. G. Bawendi, D. Fukumura, L. L. Munn and R. K. Jain, *Biophys. J.,* 2010, **5**, 1342–1349.
52. K. B. Kosto and W. M. Deen, *AIChE J.,* 2004, **50**, 2648–2658.
53. A. D. Lander, Q. Nie, B. Vargas and F. Y. M. Wan, *SIAM J. Appl. Dyn. Syst.,* 2005, **114**, 343–374.
54. R. L. Fournier, in *Basic Transport Phenomena in Biomedical Engineering,* Taylor & Francis, New York, 2nd edn, 2007, ch. 5.
55. D. S. Clague and R. J. Phillips, *Phys. Fluids,* 1996, **8**, 1720–1731.
56. A. Pluen, Y. Boucher, S. Ramanujan, T. D. McKee, T. Gohongi, E. Tomaso, E. B. Brown, Y. Izumi, R. B. Campbell, D. A. Berk and R. K. Jain, *Proc. Natl. Acad. Sci. U. S. A.,* 2001, **98**, 4628–4633.
57. I. F. Tannock, C. M. Lee, J. K. Tunggal, D. S. M. Cowan and M. J. Egorin, *Clin. Cancer Res.,* 2002, **8**, 878–884.
58. M. Magzoub, J. Songwan and A. S. Verkman, *FASEB J.,* 2008, **22**, 276–284.
59. B. J. Peret and W. L. Murphy, *Adv. Funct. Mater.,* 2008, **18**, 3410–3417.
60. G. M. Cruise, D. S. Scharp and J. A. Hubbell, *Biomaterials,* 1998, **19**, 1287–1294.
61. E. H. Nguyen, M. P. Schwartz and W. L. Murphy, *Macromol. Biosci.,* 2011, **11**, 483–492.
62. S. R. Lustig and N. A. Peppas, *J. Appl. Polym. Sci.,* 1988, **36**, 735–747.
63. J. M. Savinell, G. M. Lee and B. O. Palsson, *Bioprocess Biosyst. Eng.,* 1989, **4**, 231–234.
64. K. Francis and B. O. Palsson, *Proc. Natl. Acad. Sci. U. S. A.,* 1997, **94**, 12258–12262.
65. W. M. Saltzman, in *Tissue Engineering: Principles for the Design of Replacement Organs and Tissues*, Oxford University Press, New York, 2004, ch. 3, 11 and Appendix B.
66. E. P. Widmaier, H. Raff and K. T. Strang, in *Vander's Human Physiology*, McGraw-Hill, New York, 11th edn, 2006, ch. 1, 3–5.

67. H. Lodish, A. Berk, C. A. Kaiser, M. Krieger, M. P. Scott, A. Bretscher, H. Ploegh, and P. Matudaira, in *Molecular Biology of the Cell*, W. H. Freeman and Company, New York, 6th edn, 2008, ch. 10–15.
68. F. T. H. Wu, M. O. Stefanini, F. M. Gahann and A. S. Popel, *Methods Enzymol.*, 2009, **467**, 461–497.
69. D. A. Lauffenburger and J. Linderman, in *Receptors: Models for Binding, Trafficking, and Signaling*, Oxford University Press, New York, 1993, ch. 1 and 2.
70. B. V. Slaughter, S. S. Khurshid, O. Z. Fisher, A. Khademhosseini and N. A. Peppas, *Adv. Mater.*, 2009, **21**, 3307–3329.
71. R. B. Bird, W. E. Stewart and E. N. Lightfoot, *Transport Phenomena*, John Wiley & Sons, New York, 2nd edn, 2007, ch. 17–24.
72. J. Taipale and J. Keski-oja, *FASEB J.*, 1997, **11**, 51–59.
73. M. Princivalle and A. Agostini, *Int. J. Dev. Biol.*, 2002, **46**, 267–278.
74. I. Vlodavsky, R. Bar-Shavit, R. Ishar-Michael, P. Bashkin and Z. Fuks, *Trends Biochem. Sci.*, 1991, **16**, 268–271.
75. R. G. Thorne, A. Lakkaraju, E. Rodriguez-Boulan and C. Nicholson, *Proc. Natl. Acad. Sci. U. S. A.*, 2008, **105**, 8416–8421.
76. S. P. Zustiak, H. Boukari and J. B. Leach, *Soft Matter*, 2010, **6**, 3609–3618.
77. T. Nie, A. Baldwin, N. Yamaguchi and K. L. Kiick, *J. Controlled Release*, 2007, **122**, 287–296.
78. T. Nie, R. E. Akins and K. L. Kiick, *Acta Biomater.*, 2009, **5**, 865–875.
79. K. L. Kiick, *Soft Matter*, 2008, **1**, 29–37.
80. R. Jin, L. S. T. Moreira, A. Krouwels, P. J. Dijkstra, C. A. Blitterswijk, M. Karperien and J. Feijen, *Acta Biomater.*, 2010, **6**, 1968–1977.
81. S. Kondo and T. Miura, *Science*, 2010, **329**, 1616–1620.
82. P. K. Kythe, in *Fundamental Solutions of Partial Differential Equations and Applications*, Birkhauser, New Orleans, 1996.
83. M. A. Rice, J. Sanchez-Adams and K. S. Anseth, *Biomacromolecules*, 2006, **7**, 1968–1975.
84. H. Nagase and G. B. Fields, *Biopolymers*, 1996, **40**, 399–416.
85. M. E. Fleury, K. C. Boardman and M. A. Swartz, *Biophys. J.*, 2006, **91**, 113–121.
86. H. C. Brinkman, *Appl. Sci. Res., Sect. A*, 1947, **1**, 27–34.
87. G. E. Kapellos, T. S. Alexiou and A. C. Payatakes, *Math. Biosci.*, 2010, **225**, 83–93.
88. G. W. Jackson and D. F. James, *Can. J. Chem. Eng.*, 1986, **64**, 364–374.
89. V. Kapur, J. C. Charkoudian, S. B. Kessler and J. L. Anderson, *Ind. Eng. Chem. Res.*, 1996, **35**, 3179–3185.
90. W. Lin, K. R. Shull, C. Hui and Y. Lin, *J. Chem. Phys.*, 2007, **127**, 094906.

CHAPTER 2

Biocompatibility of Hydrogelators Based on Small Peptide Derivatives

YI KUANG, NING ZHOU, AND BING XU*

Brandeis University, 415 South Street, Waltham, MA 02453, USA
*E-mail: bxu@brandeis.edu

2.1 Introduction

Hydrogels, composed of elastic networks that immobilize up to 99% of water, first introduced and classified by Hardy in 1900,[1] are now becoming an important class of biomaterials and have attracted growing interest in the past two decades.[2] The merits of hydrogels lie not only in their highly hydrated nature, which provides a semi-wet environment, but also, in many cases, stem from the inert surfaces provided by their elastic networks.[3] Usually, chemical crosslinking or physical self-assembly of synthetic or natural building blocks, a step called "hydrogelation", results in the hydrogels. Because chemically crosslinked hydrogels have limitations such as slow biodegradation, a restricted number of hydrogelation methods,[4] the development of the physically self-assembled hydrogels, as alternatives of chemically crosslinked polymeric hydrogels, has received increased exploration in recent years. Because non-covalent supramolecular interactions[5,6] drive the spontaneous arrangement of molecules (*i.e.*, hydrogelators) in them, the physically self-assembled hydrogels are also referred as supramolecular hydrogels.

There are several strategies for generating supramolecular hydrogels: the use of biological building blocks, such as polysaccharides (chitin,[7] agarose[8]), matrix proteins (collagens,[9] elastins[10]), or other proteins or oligopeptides that self-assemble in water,[11] the design of synthetic molecules,[12,13] and the combination of biological and synthetic build blocks. One useful concept in the last strategy is to conjugate small peptides with synthetic (organic or inorganic) molecules.[14] There are several advantages for using small peptides. First, the native structure of amino acids and peptides favours self-assembly. The extensive hydrogen-bond donors and acceptors forming the peptide backbone are a key factor that dictates the formation of ordered secondary structure of peptides.[15] Amino acids with an aromatic ring also offer efficient intermolecular aromatic–aromatic interactions. Second, it is easy to tune the hydrophobicity and hydrophilicity of peptides by selecting a proper combination of amino acid residues. Third, small peptide building blocks are highly versatile in both structure and function (protein-binding peptides, peptidase cleavage site, ion-binding site, hormones,[16] antibiotics,[17] toxin,[18] *etc.*) that enable a diverse design of hydrogelators and provide the resulting hydrogels with a wide range of biological activities. Fourth, compared to proteins and oligopeptides, small peptides have better biodegradability[19] and are synthetically feasible and economic.

The strategy of the conjugation of small peptides with synthetic (organic or inorganic) molecules not only results in supramolecular hydrogels, but also confers new capabilities to the building blocks of the hydrogelators. A motif that is prone to self-assemble can facilitate the hydrogelation of a functional small peptide that is unable to self-assemble by itself. In this way, the hydrogels preserve the biological epitopes and corresponding functions of the functional peptides.[20] On the other hand, a self-assembling peptide can conjugate with organic or inorganic molecules that are unable to self-assemble, which allows the materials to exhibit the functionalities of the synthetic molecules.[21] Because of their great potential, hydrogelators based on small peptide derivatives and their hydrogels are being explored extensively. For example, a diverse range of micro- or nanoscale architectures (*e.g.* fibres, ribbons, tubes, sheets, and spheres) formed by small peptide derivatives have served as the networks of hydrogels.[22] The hydrogels can also have a range of different physical properties, including electrical,[23] optical,[24] and mechanical properties.[25] Besides having interesting physical properties, the hydrogelators based on small peptide derivatives have great potential as biomaterials. Pioneering studies have demonstrated that the hydrogels of small peptide derivatives are able to serve as potential biomaterials with a wide range of functions and applications ranging from drug delivery,[26] tissue engineering,[4] neuron regeneration,[27] bioactive platforms,[28,29] and biomineralization,[30] to cell culturing.[31]

Although the above studies have largely inspired the exploration of hydrogels formed by hydrogelators based on small peptide derivatives, it is necessary to evaluate first the biocompatibility of the hydrogelators before using the corresponding hydrogels as biomaterials. Obviously, the cytotoxicity or biocompatibility of the hydrogelators will ultimately decide the applications of the hydrogels. In this chapter, we focus on the biocompatibility of hydrogelators of small peptide derivatives according to several recent studies. The chapter is organized in the following way: we first introduce studies that explore the biocompatibility of the hydrogelators of small peptide derivatives on mammalian cells, and then summarize several works on the biocompatibility of the hydrogelators on animal models. By illustrating the biocompatibility of various types of hydrogelators containing small peptides, we hope this chapter will provide the basic framework for the design of hydrogelators of small peptide derivatives that aim towards biomedical applications.

2.2 Biocompatibility of Hydrogelators in Mammalian Cells

Regardless of their specific biomedical applications, supramolecular hydrogels usually have to contact or enter mammalian cells to achieve their functions. Therefore, it is critical to assess the biocompatibility of the hydrogelators made of small peptide derivatives in mammalian cells. In this section, we discuss biocompatibility of several representative hydrogelators. These hydrogelators all contain a short peptidic segment, but connect to different motifs at the C-terminus, N-terminus, or both ends of peptides. First, we discuss the fact that the hydrogelators consist of different aromatic motifs at the N-terminals of pentapeptides to exhibit different degree of cytotoxicity. Second, we introduce the hydrogelators of N-unsubstituted peptides that display cell-specific cytotoxicity. Third, we discuss the biocompatibility of hydrogelators made of peptides that conjugate with nucleobases at the N-terminus of the peptides. Finally, we demonstrate the incorporation of nucleobases and saccharides to each end of peptides to afford hydrogelators that exhibit exceptional biocompatibility.

2.2.1 Biocompatibility of Hydrogelators of Conjugates of Pentapeptides and Aromatic Motifs

Supramolecular hydrogelators require a subtle balance of hydrophobicity/hydrophobicity and adequate amount of intermolecular interactions to self-assemble into a hydrogel. Therefore, a slight change of the molecular structure might induce a drastic difference in their ability to self-assemble. Due to the difficulty of correlating the molecular structure with the self-assembly properties, it still remains a challenge to predict whether a small peptide derivative can act as a hydrogelator. However, several representative hydrogelators all share a common structural feature—they contain aromatic moieties. The incorporation of aromatic moieties generates critical intermolecular aromatic–aromatic interactions that serve as a strong stabilizing force for molecular self-assembly in water. Thus, aromatic moieties are one type of important motif for promoting the ability of small peptides to self-assemble. In order to correlate the cytotoxicity of hydrogelators containing aromatic moieties to their aromatic moiety, we synthesized a series of hydrogelators that are the conjugates of pentapeptides and aromatic moieties and evaluated the cytotoxicities of these hydrogelators or conjugates.[32]

Figure 2.1 shows the chemical structures of the pentapeptidic epitopes and aromatic moieties used to construct the hydrogelators and the optical images of the corresponding hydrogels. Although the five pentapeptides, which are functional epitopes of certain proteins, are unable to self-assemble in water, the covalent attachment of pyrene (**P**), fluorene (**F**), or naphthalene (**N**), the moiety that enhances aromatic–aromatic interactions, to the N-terminal of the pentapeptides results in several supramolecular hydrogelators that form

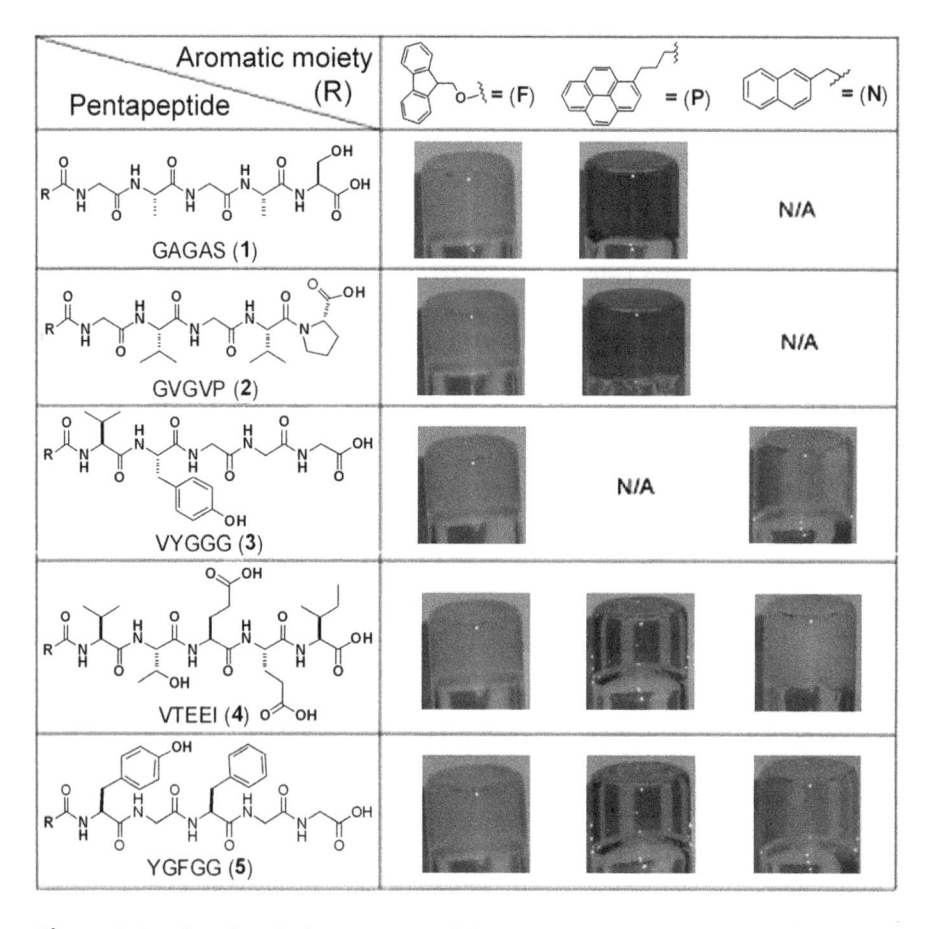

Figure 2.1 The chemical structures of the pentapeptidic epitopes and aromatic moieties used for constructing hydrogelators and the optical images of the hydrogels formed by some of the resulting conjugates (N/A indicates no hydrogelation). Adapted from ref. 32.

hydrogels on change of pH. Specifically, GAGAS (**1**) and GVGVP (**2**) conjugate with naphthalene to generate **1-N** and **2-N**, which are highly hydrophilic and unable to form hydrogels. But the conjugation of GAGAS and GVGVP with more hydrophobic and larger aromatic moieties, pyrene and fluorene, results in hydrogelators **1-P** and **2-P** that form hydrogels at 1 wt% and **1-F** and **2-F** that form hydrogels at 0.5 wt%, respectively. VYGGG conjugates with naphthalene and fluorene moieties to result in hydrogelators **3-N** and **3-F**, but pyrene-conjugated VYGGG (**3-P**) is so hydrophobic that it precipitates from water under all pH conditions. Unlike the above three pentapeptides, the conjugates of pyrene (**P**), fluorene (**F**), or naphthalene (**N**) with VTEEI and YGFGG are hydrogelators. The hydrogelation properties of these conjugates agree with the notion that adequate aromatic–aromatic interactions promote the self-assembly of small peptidic derivatives in water.

According to MTT assays on HeLa cells treated with the conjugates of pentapeptides and aromatic moieties shown in Table 2.1, all five pentapeptidic epitopes themselves have IC$_{50}$ higher than the highest concentration tested (1.5 mM, Table 2.1), suggesting that the pentapeptidic epitopes are biocompatible. Among the conjugates containing the GAGAS epitope, although **1-F** affords hydrogel and **1-N** cannot, they have a similar IC$_{50}$. But **1-P** has a comparably larger IC$_{50}$ value, indicating **1-P** has lower toxicity than **1-F** and **1-N**. For the GVGVP-based derivatives, hydrogelators **2-F** and **2-N** also have close IC$_{50}$ values despite their different hydrogelation abilities. Unlike GAGAS-based derivatives, **2-P** has much higher cytotoxicity than **2-F** and **2-N**. The derivatives of VYGGG, unlike those of GAGAS and GVGVP, regardless of their self-assembly properties, have IC$_{50}$ values higher than 1.5 mM. The VTEEI- and YGFGG-based derivatives share a similar trend: the fluorene and naphthalene conjugates have higher IC$_{50}$ values than the pyrene conjugates do. These results, however, appear to be insufficient to generalize a trend for the cytotoxicity of these supramolecular hydrogelators.

Interestingly, the cytotoxicity of pyrene-conjugated pentapeptides likely correlates with the dimerization of the pyrene group of the conjugates.[32] Figure 2.2 shows a plot of IC$_{50}$ values of the conjugates containing pyrene and their emissions at 393 nm, which is the emission peak of the monomeric forms of pyrene. Lower intensity of the emission at 393 nm indicates higher dimerization of pyrene groups. **3-P** has the lowest IC$_{50}$ value of all pyrene

Table 2.1 Summary of the IC$_{50}$ values of pentapeptides and their derivatives on HeLa cells after 48 h incubation.[a]

	Pentapeptide	*-F*	*-P*	*-N*
1	>1.5 mM	190 ± 2 µM	300 ± 4 µM	193 ± 3 µM
2	>1.5 mM	>1.5 mM	229 ± 3 µM	>1.5 mM
3	>1.5 mM	>1.5 mM	>1.5 mM	>1.5 mM
4	>1.5 mM	400 ± 2 µM	160 ± 4 µM	350 ± 2 µM
5	>1.5 mM	655 µM	231 ± 2 µM	445 ± 4 µM

[a] Adapted from ref. 32.

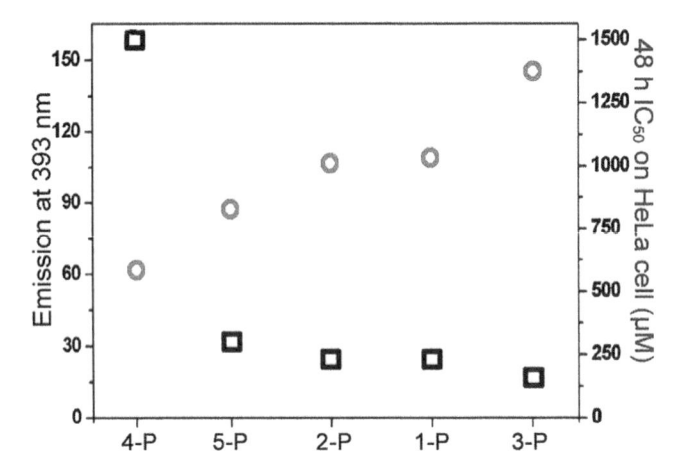

Figure 2.2 Plots of the IC$_{50}$ values (open circles) and the fluorescence intensity at 393 nm (open squares) of five pyrene conjugates of pentapeptides. Adapted from ref. 32.

conjugates, and it has the highest degree of dimerization. On the other hand, the presence of bulky side chains on VTEEI, three carboxylic acid groups and one hydroxyl group, prevents the dimerization of **4-P**, which results in the lowest degree of dimerization of the pyrene group. **4-P**, therefore, also exhibits the lowest IC$_{50}$ value.

The biocompatibilities of the hydrogelators in this study illustrate that the cytotoxicity, at least, of the pentapeptidic conjugates has little correlation with either the sequences of the pentapeptides or the ring size of the aromatic moieties. Moreover, the cytotoxicity of these pentapeptidic derivatives also has little dependence on their hydrogelation ability alone. Although there is a certain connection between the dimerization of pyrene groups of the conjugates and the cytotoxicity of conjugates, the dimerization of pyrene groups is difficult to predict simply from the structure of the conjugates. These results demonstrate that the cytotoxicity of hydrogelators is rather hard to predict, which underscores the importance of the experimental evaluation of the biocompatibility of hydrogelators before committing them for biological applications.

2.2.2 Biocompatibility of Hydrogelators of N-Unsubstituted Dipeptides

Most hydrogelators based on small peptide derivatives explored in the last decade have synthetic motifs covalently attached at the N-terminal (amine end) of the peptides, which is synthetically more feasible than those conjugates that attach a non-amino acid motif at the C-terminal (carboxylic end).[33] Hydrogelators consisting of peptides with unsubstituted N-termini (N-unsubstituted), although less explored, are important because N-unsubstituted

Biocompatibility of Hydrogelators Based on Small Peptide Derivatives 37

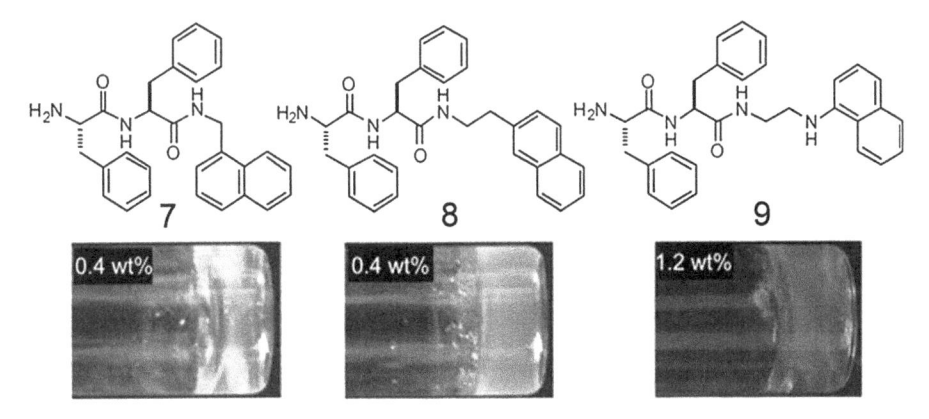

Figure 2.3 The chemical structures the hydrogelators of N-unsubstituted peptides and the optical images of the corresponding hydrogels. Adapted from ref. 34.

hydrogelators carry a different charge compared to N-substituted hydrogelators. An unsubstituted peptide C-terminal gives a negative charge to the hydrogelator, whereas an unsubstituted peptide N-terminal is likely to result in a positive charge to the hydrogelator under physiological conditions. The state of charge is one of the key factors that dictate intermolecular interaction among peptides or proteins. Thus, understanding the less explored, N-unsubstituted, hydrogelators, especially their biocompatibility, is significant for the exploration of hydrogelators derived from small peptide derivatives.

We have designed three C-terminated hydrogelators, all of which consist of a diphenylalanine and a naphthalene group.[34] Besides all having an unsubstituted N-terminal and substituted C-terminal, these three hydrogelators are structurally different in their linkers between the diphenylalanine and naphthalene groups and the positions of the substitution of naphthalene. Figure 2.3 shows that hydrogelators 7 and 9 both have 1-substituted naphthalene, while 8 has 2-substituted naphthalene. Unlike the linkers in 7 and 8, the linker of 9 has an extra secondary amine right next to the naphthalene ring. Hydrogelation tests show that all three hydrogelators can produce hydrogels at pH 5.0–6.0. Hydrogelator 9 requires higher concentration than the other two hydrogelators to form the hydrogel, likely because of the presence of the extra secondary amine.

We used MTT assays to examine the biocompatibility of the three hydrogelators in two cancer cell lines, HeLa and T98G, and a counterpart cell line of HeLa cell, Ect1/E6E7 (an immortalized normal cell line). Table 2.2 shows the IC_{50} values of the three hydrogelators. Although hydrogelator 8 has an IC_{50} greater than 200 µM, hydrogelators 7 and 9 exhibit considerable cytotoxicity (IC_{50} value about 150 µM) towards HeLa cells. All the three hydrogelators also exhibit cytotoxicity towards T98G cells, indicating that these N-unsubstituted hydrogelators are generally cytotoxic towards cancer cell

Table 2.2 Summary of the IC$_{50}$ values of N-unsubstituted hydrogelators on different cell lines after 48 h incubation.[a]

Hydrogelator	IC$_{50}$ (µM)		
	HeLa	T98G	Ect1/E6E7
7	142.0 ± 3.6	112.9 ± 6.8	>200
8	>200	150.4 ± 6.6	>200
9	150.2 ± 3.1	104.4 ± 6.3	>200

[a] Adapted from ref. 34.

lines. However, the three hydrogelators all have IC$_{50}$ value higher than 200 µM in Ect1/E6E7 cells, demonstrating that these N-unsubstituted hydrogelators are less toxic to Ect1/E6E7 cells than to cancer cells. Moreover, in both HeLa and T98G cell lines, **8** has higher IC$_{50}$ values than **7** and **9** do. Since **8** has 2-substituted naphthalene and **7** and **9** have 1-substituted naphthalenes, this result suggests that the hydrogelators containing 2-substituted naphthalene might be less toxic than those containing 1-substituted naphthalene.

Selective cytotoxicity to cancer cells is a useful property of N-unsubstituted hydrogelators, which differentiates the N-unsubstituted hydrogelators from N-substituted hydrogelators. This finding provides a new insight for the design of supramolecular hydrogels as biomaterials for cancer therapeutics, a research direction that is receiving increased attention.[35–37] Moreover, the correlation of the substitution on the naphthalene group with the cytotoxicity of the hydrogelators provides subtle and previously unknown information for the design of small peptidic hydrogelators to achieve desired biocompatibility or cytotoxicity.

2.2.3 Biocompatibility of Hydrogelators of the Conjugates of Nucleobases and Peptides

The integration of nucleobases with small peptides provides an opportunity to generate nucleopeptides that can carry a large variety of biological functions. Besides, since both nucleobases and amino acids are naturally occurring building blocks, conjugates made of these two types of building blocks likely will be biodegradable and biocompatible. Thus, it is no surprise that the biological functionalities and biocompatibility attract interest in conjugating nucleobases with small peptides to construct supramolecular hydrogelators. Figure 2.4 shows two novel designs of hydrogelators formed by the conjugation of small peptides and nucleobases.[38] In design I (**10–A, 10–T, 10–C,** and **10–G**), the nucleobase covalently links to the N-terminal of a diphenylalanine. In design II (**11–A, 11–T, 11–C,** and **11–G**), the nucleobase conjugates to the lysine side chain of a tripeptidic derivative that contains a naphthalene group at the N-terminal to increase the hydrophobicity and intermolecular aromatic interactions. All eight molecules yield hydrogels at

Figure 2.4 Chemical structures of hydrogelators of the conjugates of nucleobases and peptides.

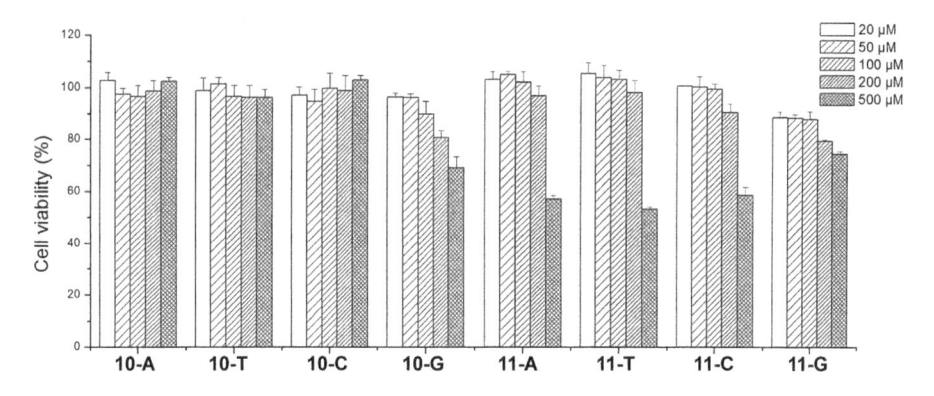

Figure 2.5 The cytotoxicity of the hydrogelators of the nucleopeptides towards HeLa cells. Adapted from ref. 38.

2 wt%, indicating that the conjugation of nucleobases and small peptides is a valid route for constructing supramolecular hydrogelators.

To verify the biocompatibility of the hydrogelators, MTT assays were used to measure the viability of HeLa cells treated with the hydrogelators of the nucleopeptides. Figure 2.5 shows the percentage viability of the HeLa cells treated with the hydrogelators for 72 h. Even at the highest concentration tested, **10-A, 10-T,** and **10-C** induce little viability loss of the HeLa cells. However, cell viability decreases slightly when treated with 500 µM of **10-G**, but over 50% of HeLa cells are still viable (IC$_{50}$ of **10-G** >500 µM). On the other hand, while 500 µM of **11-A, 11-T,** or **11-C** all decrease cell viability to about 50%, 500 µM of **11–G** decreases cell viability by less than 30%, indicating that **11–G** is less toxic than the other three. In both design I and design II, guanine-containing hydrogelators exhibit different toxicity from the hydrogelators containing the other three nucleobases. The exception of guanine-containing hydrogelators likely stems from the unique ability of guanine to form a G-quadruplex.[39] **10–A**, **10–T**, and **10–C** are all more biocompatible than **10–G**, demonstrating that the attachment of nucleobases to N-terminal of small peptides results in biocompatible hydrogelators. On the contrary, the conjugation of nucleobases to the side chain of a tripeptide derivative affords hydrogelators that exhibit moderate cytotoxicity, likely due to the incorporation of the naphthalene group.

The results of the hydrogelators in design I suggest that the conjugates of nucleobases and small peptides are highly biocompatible, thus the hydrogelators of nucleopeptides should provide the corresponding hydrogels as functional biomaterials. Moreover, this study points out that guanine might possess properties different from the other three nucleobases, so guanine-containing hydrogelators likely will need careful examination on biocompatibility.

2.2.4 Biocompatibility of Hydrogelators of Conjugates of Nucleobases, Peptides, and Saccharides

Nucleotides, amino acids, and saccharides are three of the four basic building blocks of life. While water accounts for about 70% of a cell's weight, nucleotides, amino acids, and saccharides, together with fatty acids, constitute the majority of the remaining 30% of cell mass. Besides using one of the three building blocks to build macromolecules, such as polysaccharides and proteins, nature also combines more than one type of building block to construct macromolecules. For example, glycoproteins consist of covalently attached saccharides and amino acids[40] and tRNA carries an amino acid on the 3′ end of the nucleotide sequence.[41] Inspired by the elegance of tRNA, we synthesized a class of hydrogelators that are abiotic conjugates of nucleobases, peptides, and saccharides.[42]

Figure 2.6 shows the chemical structures of these hydrogelators. Nucleobases were covalently attached to the N-terminal of small peptide phenylalanine (design III) or diphenylalanine (design IV), and a D-glucosamine was attached to the C-terminal of the peptide. Considering the fundamental importance of these three motifs in biology, we described this kind of conjugate as a "molecular trinity".[42] Designs III and IV, together, result in eight conjugates, and all of them are able to form hydrogels at a concentration of about 3 wt% in water and at neutral pH.

MTT cell viability tests show that conjugates based on the "molecular trinity" exhibit excellent biocompatibility. Figure 2.7 shows the viability of HeLa cells treated with the hydrogelators in Figure 2.6 for 72 h. At the highest concentration treated by the hydrogelators, over 80% of HeLa cells are still viable, which indicates low cytotoxicity of all the hydrogelators in

Figure 2.6 Chemical structures of hydrogelators of the conjugates of nucleobases, peptides, and saccharides.

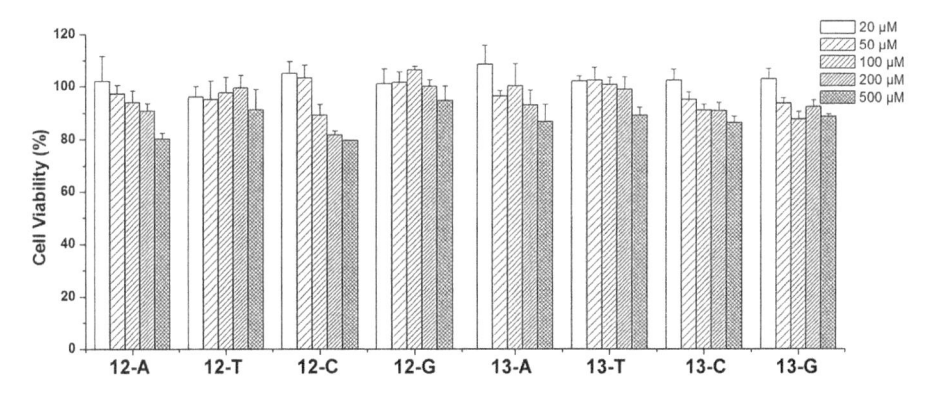

Figure 2.7 Cytotoxicity of hydrogelators of the conjugates of nucleobases, peptides, and saccharides on HeLa cells. Adapted from ref. 42.

Figure 2.6. Especially notable, **13–G** is more biocompatible than **10–G**. Because the only structural difference between **13–G** and **10–G** is the glucosamine at the C-terminal, this result suggests that the incorporation of saccharide may be a useful approach to reduce the cytotoxicity of hydrogelators of nucleopeptides.

This study suggests that the simple conjugation of nucleobases, amino acids, and saccharides offers an effective way to generate a large pool of highly biocompatible supramolecular hydrogelators, which likely will provide a variety of biocompatible hydrogels for expanding the current exploration of biofunctional supramolecular hydrogels.

2.3 Biocompatibility of Hydrogelators in Animal Models

Cells are the basic structural and functional unit in all living organisms.[43] Because animals consist of various types of cells that are hierarchically organized, they are complex organisms and differ from single cells. Therefore, to validate hydrogelators as suitable biomaterials, after testing biocompatibility of hydrogelators in cell assays, the next necessary and logical step is to evaluate the biocompatibility of hydrogelators in animal models. Depending on the application, there are different tests for *in vivo* biocompatibility.[44] For example, systemic toxicity tests measure acute toxicity of compounds on animals;[45] haemocompatibility tests, such as the thrombosis assay, measure compatibility of materials that comes into contact with blood.[46] The most frequently used method for testing biocompatibility of hydrogelators in animals is immune response evaluation, which assesses immunogenicity of non-natural materials.[44]

In a typical immune response evaluation, animals are administered a certain amount of hydrogelators under their skin (by subcutaneous injection). After a certain length of time (*e.g.* days), histological staining of the skin at the injection site, by haematoxylin and eosin (H&E) reveals the existence of immune cells in the tissue. Normal tissue has a only small amount of immune cells, indicating high biocompatibility, and an accumulation of immune cells indicates low biocompatibility. Using this method, we examined the biocompatibility of a hydrogelator (**14**), which is a D-peptidic derivative and a candidate for drug delivery, on SD rats.[47] Figure 2.8 shows the chemical structure of the hydrogelator **14** and the H&E stained skin of SD rats after subcutaneous injection of **14**. H&E staining shows that the skin tissues of the rats with and without injection of **14** have little difference. There are few immune cells which stain strongly by H&E staining (red for acidophilic cytoplasm, blue for basophil nuclei, and red for erythrocytes). The immune response evaluation suggests that hydrogelator **14** is biocompatible in these animals.

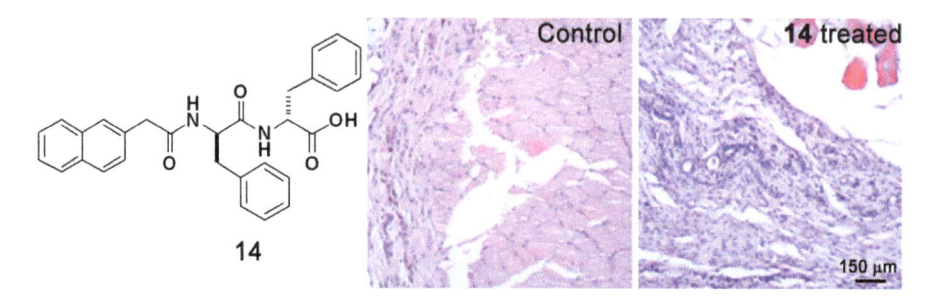

Figure 2.8 Chemical structure of hydrogelator **14** and the histological cross-section images of skin of untreated rat and skin of injection spot of SD rats on day 42 after injection of **14** (0.2 mL, 0.8 wt%). Adapted from ref. 47.

Besides subcutaneous injection, a variant method for immune response evaluation is a wound healing assay. In a wound healing assay, a cut is made on the skin of the animal to recruit immune cells, and then hydrogelators are applied to the wound site.[48] The pre-recruitment of immune cells allows intensive interactions between the immune system and the hydrogelators applied to the wound. The merit of wound healing assays is that the healing of the wound reflects the result of the immune response, making it an easier and more time-efficient assay than histology and H&E staining. We used the wound healing assay to verify the biocompatibility of a glycopeptide hydrogelator that exhibits high biocompatibility in cells.[49] Figure 2.9 shows the structure of hydrogelator **15** and the images of wounds on the SD rats with and without treatment with hydrogelator **15**. As shown in Figure 2.9, the wound treated with hydrogelator **15** has a smaller scar than the untreated wound, indicating that hydrogelator **15** not only is biocompatible on animals, but also promotes wound healing. The H&E staining of histological specimens of wound treated by hydrogelator **15** reveals smaller scar tissue, which agrees with the observation that hydrogelator **15** promotes wound healing. This result also validates wound healing as an alternative assay for evaluating the biocompatibility of hydrogelators on animals.

Another method, which also measures the immune responses based on the symptom of the animal instead of the local presence of immune cells, is to monitor the weight change of the animals.[50] The idea behind of the method is that severe immune responses might induce weight gain or weight loss. We measured the weight change of mice receiving subcutaneous or intraperitoneal injection of a hydrogelator precursor **16a** (0.5 mL, 0.8 wt%), which evaluates responses of skin or blood to **16a**, respectively. Prior to animal testing, MTT assays reveal that the precursor **16a** is moderately toxic (IC$_{50}$ = 93 µM) while its hydrogelator **16b**, formed by the removal of phosphate group of **16a** catalysed by phosphatases, is biocompatible (IC$_{50}$ = 603 µM). Figure 2.10 shows the chemical structures of **16a** and **16b**, and the plot of weight gain of animals against the number of days after the injection of **16a**. As shown in the plot, the weight change of the mice receiving subcutaneous injection of saline or **16a** remained statistically the same, suggesting that subcutaneous injection of **16a** results in little acute toxicity to the mice at the experimental dosage. In mice receiving intraperitoneal injection of saline or **16a**, although there is an obvious difference in weight gain in the first day after treatment, the two groups have almost the same rate of weight gain (0.34 g per day) following the second day after treatment. These results suggest that intraperitoneal injection of **16a** results in slight acute toxicity, which likely originates from the toxicity of **16a**, but conversion of **16a** to the less toxic **16b** diminishes the toxicity induced by **16a**. The slight difference in response to **16a** indicates same hydrogelator might have distinct responses to different tissues, which highlights the importance of thorough assessment of the biocompatibility of hydrogelators in animals.

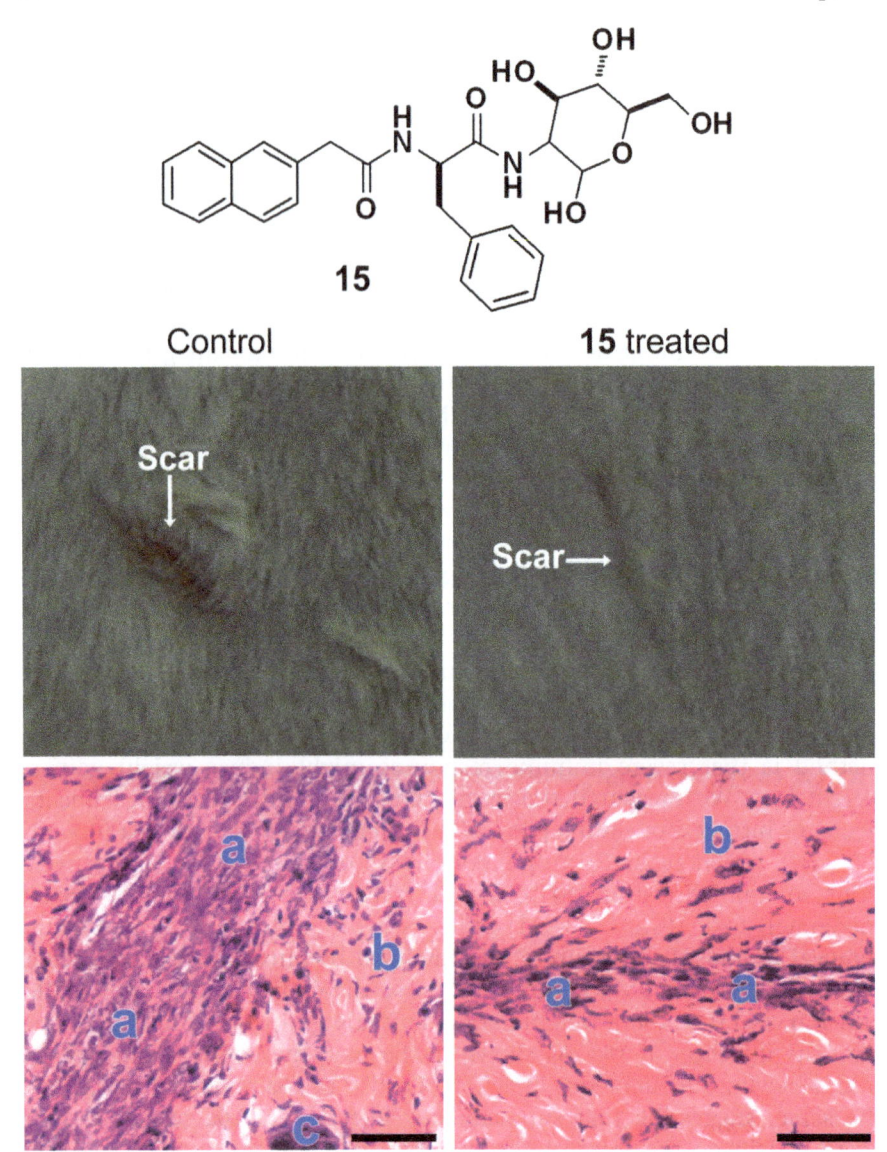

Figure 2.9 Chemical structure of hydrogelator **15** and the pictures of wound spot on the SD rats and H&E staining of the wound spot after 18 days of treatment with and without hydrogelator **15** (1 mL, 0.2 wt%). In the H&E staining images, a represents scar tissue; b represents extracellular matrix, and c represents keratinocytes. Scale bar = 50 μm. Adapted from ref. 49.

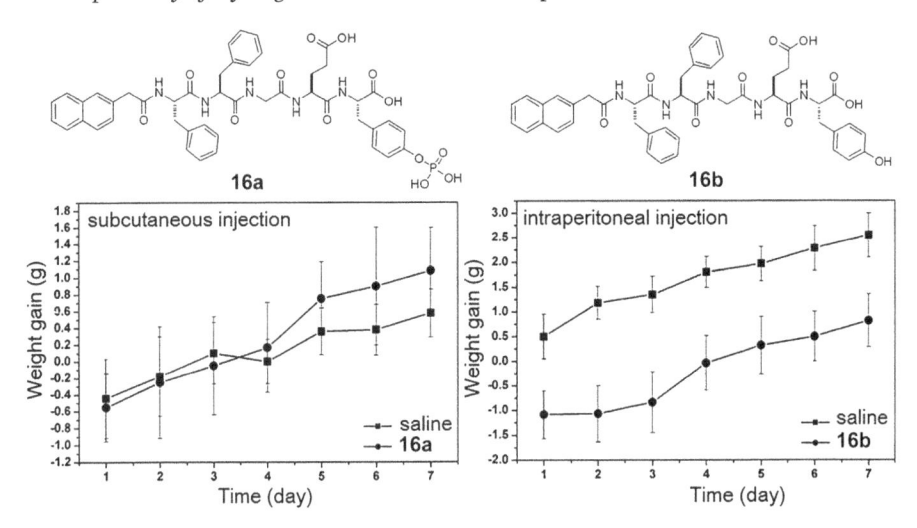

Figure 2.10 Chemical structures of hydrogelator precursor **16a** and the corresponding hydrogelator **16b**. Plot of time against weight gain of mice received subcutaneous or intraperitoneal injection of **16a** or saline as control ($n = 6$). Adapted from ref. 50.

2.4 Conclusions

Using a variety of hydrogelators containing small peptides, we carried out biocompatibility studies of these hydrogelators in mammalian cells and in animals. The biocompatibility tests in mammalian cells imply several plausible trends that may correlate the structures and the biocompatibility of the hydrogelators. While these trends may serve as a qualitative guide for the design of hydrogelators that contain small peptides to achieve certain functions without compromising the desired biocompatibility or cytotoxicity, the biocompatibility or cytotoxicity of a hydrogelator still largely depends on its overall structure. Therefore, cell and animal assays remain as necessary tests for profiling the biocompatibility of supramolecular hydrogelators. With the increased development of hydrogelators and hydrogels that contain peptide motifs, evaluation of the biocompatibility of these materials in cells and animals will contribute to expand both the knowledge and the biomedical applications of peptide-based biomaterials.

References

1. W. B. Hardy, *Z. Phys. Chem.*, 1900, **33**, 579.
2. B. Xu, *Langmuir*, 2009, **25**, 8375.
3. R. V. Ulijn, N. Bibi, V. Jayawarna, P. D. Thornton, S. J. Todd, R. J. Mart, A. M. Smith and J. E. Gough, *Mater. Today*, 2007, **10**, 40.
4. D. J. Adams and P. D. Topham, *Soft Matter*, 2010, **6**, 3707.

5. E. A. Appel, J. del Barrio, X. J. Loh and O. A. Scherman, *Chem. Soc. Rev.*, 2012, **41**, 6195.
6. M. de Loos, B. L. Feringa and J. H. van Esch, *Eur. J. Org. Chem.*, 2005, 3615.
7. J. K. F. Suh and H. W. T. Matthew, *Biomaterials*, 2000, **21**, 2589.
8. S. Hjerten, *Biochim. Biophys. Acta*, 1962, **62**, 445.
9. J. A. Fallas, L. E. R. O'Leary and J. D. Hartgerink, *Chem. Soc. Rev.*, 2010, **39**, 3510.
10. S. R. MacEwan and A. Chilkoti, *Biopolymers*, 2010, **94**, 60.
11. M. R. Caplan, P. N. Moore, S. G. Zhang, R. D. Kamm and D. A. Lauffenburger, *Biomacromolecules*, 2000, **1**, 627.
12. B. G. Xing, C. W. Yu, K. H. Chow, P. L. Ho, D. G. Fu and B. Xu, *J. Am. Chem. Soc.*, 2002, **124**, 14846.
13. D. Haldar, *Tetrahedron*, 2008, **64**, 186.
14. M. Zelzer and R. V. Ulijn, *Chem. Soc. Rev.*, 2010, **39**, 3351.
15. F. H. C. Crick, *Acta Crystallogr.*, 1953, **6**, 689.
16. H. Croxatto, W. Badia and R. Croxatto, *Proc. Soc. Exp. Biol. Med.*, 1948, **69**, 422.
17. H. S. Won, S. J. Jung, H. E. Kim, M. D. Seo and B. J. Lee, *J. Biol. Chem.*, 2004, **279**, 14784.
18. S. Mouhat, B. Jouirou, A. Mosbah, M. De Waard and J. M. Sabatier, *Biochem. J.*, 2004, **378**, 717.
19. X. M. Li, X. W. Du, J. Y. Li, Y. Gao, Y. Pan, J. F. Shi, N. Zhou and B. Xu, *Langmuir*, 2012, **28**, 13512.
20. Z. M. Yang, G. L. Liang, M. L. Ma, Y. Gao and B. Xu, *J. Mater. Chem.*, 2007, **17**, 850.
21. Z. M. Yang, Y. Kuang, X. M. Li, N. Zhou, Y. Zhang and B. Xu, *Chem. Commun.*, 2012, **48**, 9257.
22. M. Reches and E. Gazit, *Nano Lett.*, 2004, **4**, 581.
23. H. Shao and J. R. Parquette, *Chem. Commun.*, 2010, **46**, 4285.
24. J. H. Kim, S. Y. Lim, D. H. Nam, J. Ryu, S. H. Ku and C. B. Park, *Biosens. Bioelectron.*, 2011, **26**, 1860.
25. Z. M. Yang, L. Wang, J. Y. Wang, P. Gao and B. Xu, *J. Mater. Chem.*, 2010, **20**, 2128.
26. F. Zhao, M. L. Ma and B. Xu, *Chem. Soc. Rev.*, 2009, **38**, 883.
27. M. A. Bokhari, G. Akay, S. G. Zhang and M. A. Birch, *Biomaterials*, 2005, **26**, 5198.
28. N. Mari-Buye, S. O'Shaughnessy, C. Colominas, C. E. Semino, K. K. Gleason and S. Borros, *Adv. Funct. Mater.*, 2009, **19**, 1276.
29. L. S. Ferreira, S. Gerecht, J. Fuller, H. F. Shieh, G. Vunjak-Novakovic and R. Langer, *Biomaterials*, 2007, **28**, 2706.
30. J. D. Hartgerink, E. Beniash and S. I. Stupp, *Science*, 2001, **294**, 1684.
31. T. Liebmann, S. Rydholm, V. Akpe and H. Brismar, *BMC Biotechnol.*, 2007, **7**, 88.
32. M. L. Ma, Y. Kuang, Y. Gao, Y. L. Zhang, P. Gao and B. Xu, *J. Am. Chem. Soc.*, 2010, **132**, 2719.

33. Y. Zhang, Y. Kuang, Y. A. Gao and B. Xu, *Langmuir,* 2011, **27**, 529.
34. Y. Kuang, Y. Gao and B. Xu, *Chem. Commun.,* 2011, **47**, 12625.
35. R. H. Zha, S. Sur and S. I. Stupp, *Adv. Healthcare Mater.,* 2013, **2**, 126.
36. A. Altunbas, S. J. Lee, S. A. Rajasekaran, J. P. Schneider and D. J. Pochan, *Biomaterials,* 2011, **32**, 5906.
37. Y. Gao, Y. Kuang, Z. F. Guo, Z. Guo, I. J. Krauss and B. Xu, *J. Am. Chem. Soc.,* 2009, **131**, 13576.
38. X. M. Li, Y. Kuang, H. C. Lin, Y. Gao, J. F. Shi and B. Xu, *Angew. Chem., Int. Ed.,* 2011, **50**, 9365.
39. T. M. Bryan and P. Baumann, *Mol. Biotechnol.,* 2011, **49**, 198.
40. Y. Funakoshi and T. Suzuki, *Biochim. Biophys. Acta,* 2009, **1790**, 401.
41. R. W. Holley, J. Apgar, G. A. Everett, J. T. Madison, M. Marquise, S. H. Merrill, J. R. Penswick and A. Zamir, *Science,* 1965, **147**, 1462.
42. X. M. Li, Y. Kuang and B. Xu, *Soft Matter,* 2012, **8**, 2801.
43. H. Lodish, A. Berk, S. L. Zipursky, P. Matsudaira, D. Baltimore and J. Darnell, *Molecular Cell Biology*, St. Martin's Press, New York, 2000.
44. D. B. Ratner, A. S. Hoffman, F. J. Schoen and J. E. Lemons, *Biomaterials Science: An Introduction Materials in Medicine*, Academic Press, New York, 2004.
45. P. A. Botham, *Toxicol. In Vitro,* 2004, **18**, 227.
46. R. W. Colman, *Hemostasis And Thrombosis: Basic Principles and Clinical Practice*, Lippincott Williams & Wilkins, Philadelphia, 2006.
47. G. L. Liang, Z. M. Yang, R. Zhang, L. H. Li, Y. Fan, Y. Kuang, Y. Gao, T. Wang, W. W. Lu and B. Xu, *Langmuir,* 2009, **25**, 8419.
48. L. A. DiPietro and A. L. Burns, *Wound Healing: Methods and Protocols*, Humana Press, Totowa, NJ, 2003.
49. Z. M. Yang, G. L. Liang, M. L. Ma, A. S. Abbah, W. W. Lu and B. Xu, *Chem. Commun.,* 2007, 843.
50. Z. M. Yang, G. L. Liang, L. Wang and B. Xu, *J. Am. Chem. Soc.,* 2006, **128**, 3038.

CHAPTER 3

Recombinant Protein Hydrogels for Cell Injection and Transplantation

PATRICK L. BENITEZ[a] AND SARAH C. HEILSHORN*[b]

[a] Bioengineering, McCullough Building, 476 Lomita Mall, Stanford, CA, USA;
[b] Materials Science and Engineering, McCullough Building, 476 Lomita Mall, Stanford, CA, USA
*E-mail: Heilshorn@Stanford.edu

3.1 Overview

As found in nature, full-length proteins consist of a genetically specified sequence of the 20 canonical amino acids, of a defined length. This sequence of chemically diverse functional groups enables the many highly controlled interactions with other molecules found in natural proteins. Recombinant proteins can be engineered to incorporate some of nature's palette of protein functionality into hydrogels for cell delivery. Current work demonstrates how this level of molecular precision can be used to address challenges in cell therapies, such as post-implantation viability, localization, and control, *via* specified gelation mechanics and tailored bioactive domains. Leveraging recombinant technology, including protein engineering, gene synthesis, expression, and purification, materials scientists have appropriated and modified naturally occurring proteins to achieve hydrogels that combine defined gelation mechanics with specified bioactive protein chemistries. Here, we specifically review recent developments in recombinant protein hydrogels that are either inspired by native extracellular matrix proteins (*e.g.* elastin, collagen, and resilin) or designed from non-matrix peptides (*e.g.* mixing-induced two-component hydrogels). In many of these case studies, domain- and sequence-level engineering enables a broad range of biochemical activity and mechanical control *via* gelation. Despite the remaining challenges of scalability and forward-designed predictability, hydrogels made of recombinant proteins offer exciting possibilities for sophisticated delivery of therapeutic cells, including multifactorial control, native-like mechanics, and sensitivity to signals from delivered cells or host tissues.

3.2 Motivation: Cellular Control Though Tailored Protein Interactions

Once delivered, therapeutic cells have the potential to stimulate host regeneration,[1] to reconstitute tissue,[2] or to serve as vehicles for disease-addressing genes.[3] Despite extensive work showing how cell-based therapies might work, translation of such therapies besides bone marrow transplants has been limited by poor viability, localization, and post-implantation control of therapeutic cells.[4,5] Clearly, if implanted cells die, leave the site of interest, or otherwise do not follow the therapeutic plan (*e.g.* if they aberrantly proliferate or differentiate), the treatment will fail. Recombinant protein hydrogels are capable of meeting these challenges by providing, on the molecular level, (1) tissue-inspired survival and other biological cues that work through native control mechanisms and (2) specialized gelation processes that mechanically protect and localize cells. These precise interactions, between cells and the materials and among constituent polymers of the material, are enabled by the application of recombinant technology to hydrogel design and synthesis.

Similar to many traditional synthetic polymers, recombinant proteins can be processed to encapsulate therapeutic cells in hydrated networks, forming implantable or injectable hydrogels. Recombinant proteins, defined as proteins that have been expressed from engineered DNA by a heterologous host, however, have special capabilities to improve therapeutic outcomes due to the molecular specification of their chemical structure. Proteins can be thought of as polymerized sequences of the natural 20 amino acids. The chemical diversity and sequence specificity of these amino acids gives rise to tremendous structural and functional diversity in fully formed proteins. For a given protein, interactions between the amino acids and the solvent thermodynamically drive the formation of a three-dimensional (3D) structure, resulting in a spatial arrangement of chemical groups that enable specific functions.[6] Thus, through manipulation of the specific identity of a protein's sequential amino acids using recombinant technology, it is possible to create biomaterials with tailored protein interactions and specific functionalities.

For applications in cell encapsulation and delivery, the needs of the tissue physiology must guide the selection of which protein sequences are incorporated into cell-bearing hydrogels. Therapeutic cells and endogenous cells at the delivery target site have complex, dynamic molecular profiles, consisting of cell-surface receptors in addition to secreted enzymes, matrix components, and soluble factors. Each of these macromolecules has evolved the capacity to interact with specific proteins,[7] and these protein interactions impact cells directly through cell-surface signalling or indirectly through changes in the microenvironment. Since these interactions are mediated through molecular recognition of complex, 3D chemical structures, engineered proteins allow us to intervene locally in a precisely controlled manner. In this sense, protein-based materials, if carefully designed, can 'hack into', disrupt, or modulate natural protein interactions, thereby controlling delivered and host cells.

3.2.1 Extracellular Protein Interactions Relevant to Cell Delivery

Though protein interactions in nature impact cells in mechanistically diverse ways, molecular recognition and specificity is a common theme. For example, cells display a variety of cell-surface receptors, *i.e.* molecular sensors that bind specific protein ligands, that allow them to detect chemicals and gradients in their environment. Binding of protein ligands often induces intracellular signalling reactions leading to proliferation,[8] motility,[9] or differentiation (see Figure 3.1a).[10] In addition, ligand binding can also result in several biophysical consequences. For example, ligand binding to cell-surface integrin receptors commonly leads to cell spreading, which can cause cytoskeletal compression of the nucleus and altered gene expression.[11] The concentration of integrin ligands can also affect motility: low concentrations result in slippage and high concentrations can trap cells, both of which impede motility.[12] Ligands can also be presented as gradients to initiate chemotaxis, *i.e.* directional motility, if the gradient is sufficient to bias intracellular factors, such as the actin cytoskeleton.[13] Taken together, these examples highlight the critical importance of protein structure in specifying ligand–receptor binding and activation to control cell behaviour.

A different example of a sequence-specific protein interaction that influences cell physiology is binding of soluble factors to matrix proteins (see Figure 3.1b). This kind of interaction decreases the effective diffusion and degradation rate of soluble factors,[14] thereby changing the time-dependence of cell responses to soluble signals.[15] Specific matrix binding sites also bind and present soluble factors with distinct orientations that can qualitatively change how and which cell-surface receptors interact with the soluble factor.[16] A common variant of this type of interaction is the binding of heparin sulfate or other polysaccharides to extracellular matrix (ECM) proteins. These polysaccharides then act as 'depots' for the specific binding of several soluble factors.[17] In a manner similar to direct binding between the matrix and soluble factors, the heparin sulfate interaction changes the

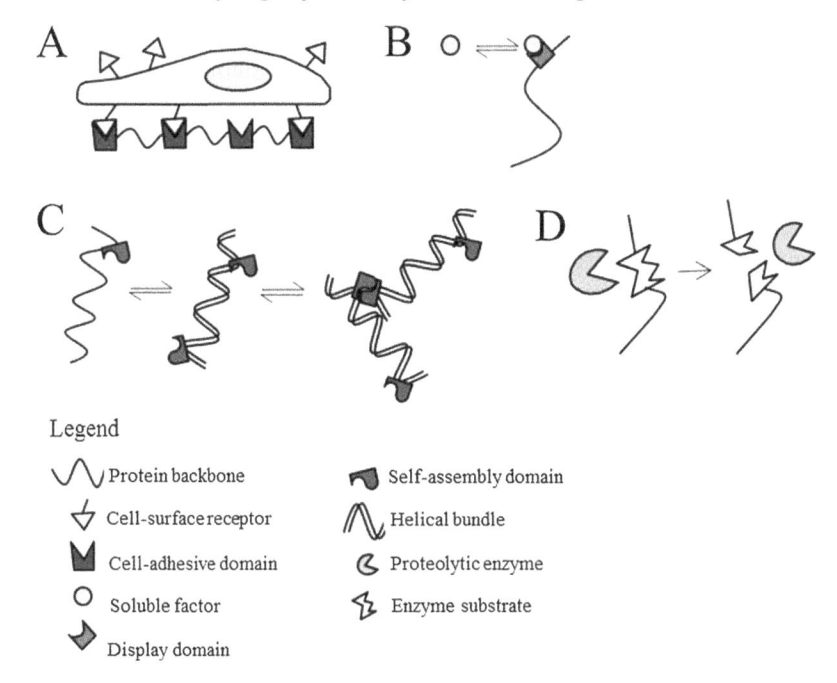

Recombinant Protein Hydrogels for Cell Injection and Transplantation

Figure 3.1 Schematic of common protein interactions found in the natural extracellular matrix, including (A) binding of extracellular ligands to cell-surface receptors, (B) immobilization of soluble factors *via* reversible binding to matrix proteins, (C) hierarchical self-assembly of matrix proteins, and (D) enzymatic proteolysis of matrix proteins at specific peptide targets.

spatiotemporal distribution and orientation of soluble factors to cell receptors.

In addition to protein interactions affecting cells directly through cell-surface receptors, many protein interactions can indirectly affect cellular behaviour. For example, protein structure and function gives rise to the mechanical properties of the ECM, which in turn alters cell morphology, migration, and differentiation.[18] Many extracellular proteins, such as fibrin and collagen, bundle or self-assemble into fibrils or more complex structures (see Figure 3.1c).[19] These ordered structures arise through specific interactions between the protein self-assembly domains. Cells themselves can manipulate this higher-order protein structure if provided with appropriate interaction domains.[20] In addition, cell-secreted enzymes, such as lysyl oxidase and transglutaminase, can stiffen extracellular protein matrices by crosslinking reactive amino acid residues.[21,22] Here, sequence specificity of the residues flanking the target amino acid determines the reaction kinetics, in addition to serving as a molecular address for a specific enzyme.[23]

Cell-secreted enzymes can also target specific amino acid sequences for proteolysis, enabling on-demand and controlled degradation of the matrix (see Figure 3.1d). This decreases the mechanical strength of the matrix and enables spatiotemporal regulation of matrix remodelling in response to specific genetic programs. For example, during wound healing programmes, matrix metalloproteinases (MMPs) are expressed that recognize specific sequences within provisional matrix proteins, allowing remodelling that spares healthy tissue.[24] In addition, proteolysis can result in the release of specific amino acid fragments, which, in turn, can be potent soluble ligands for cell receptors. For example, elastin fragments released by metalloproteinase enzymes can drive the chemotaxis of leukocytes and inflammation.[25]

As seen in the examples above, cells respond to specific protein interactions through both direct (*i.e.* ligand binding to cell-surface receptors) and indirect (*i.e.* matrix modifications that influence cell behaviour) mechanisms. Thus, during cell encapsulation and delivery, protein interactions that are designed into the biomaterial are expected to influence the therapeutic cells, the host cells, and the local microenvironment. With recombinant protein hydrogels, we seek to leverage this control to enhance the overall therapeutic efficacy of delivered cells.

3.2.2 Domain-Level Engineering in Recombinant Hydrogels

In the past, soft materials for cell delivery have been designed to engage in protein interactions with the therapeutic cells and the host tissue through various peptide conjugation chemistries.[26] While these approaches have yielded impressive results,[27,28] use of recombinant proteins to achieve molecular recognition between therapeutic cells and a protein-based cell carrier confers several potential advantages that remain insufficiently explored. One advantage of using recombinant proteins is the ability to encode full domains. In nature, proteins fold into substructures known as domains; over the course of evolution, these domains are traded, shuffled, and mutated to form new proteins with complex functions.[29] A single domain can appear in diverse proteins with diverse functions.[30] Though powerful, stepwise peptide synthesis cannot produce full-length or multidomain proteins, and this limitation motivates alternative approaches to design biologically functional medical materials, such as recombinant protein synthesis.

Full domains, though more complex, enable potent and specific interactions between engineered proteins and their targets (see Figure 3.2). Context, referring to the identity of the amino acids flanking a particular active group, influences the active group's bioactivity.[31] In the case of integrins specifically, appropriate flanking sequences are required to create a target that is recognized by specific receptors.[32] This specificity has direct consequences for which intracellular pathways are activated. Moreover, flanking residues can lead to stronger interactions, by increasing affinity or changing the kinetics of binding. Mutations in the flanking residues can alter the protein structure required for efficient binding interactions. In fact, spatially distal interactions can significantly contribute to the structural energetics of ligand

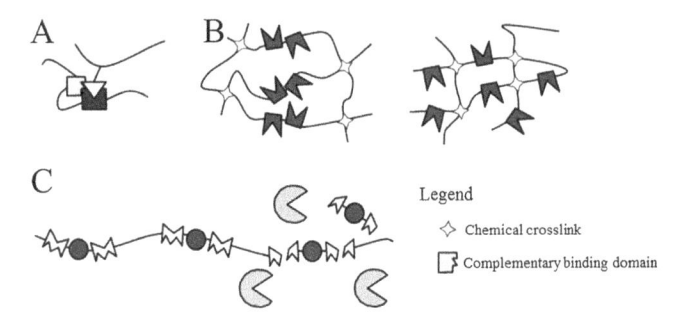

Figure 3.2 Schematic showing examples of multidomain complexity in engineered recombinant protein materials, including (A) enhanced specificity or affinity by a complementary binding domain, (B) variation of ligand spacing by selective placement of cell-binding ligands and crosslinking sites, and (C) enzyme-selective release of bioactive domains. Unlisted symbols are as described in Figure 3.1.

binding.[33] Multiple domain constructs have been engineered to take advantage of biochemical interactions between discrete domains: for example, combining two domains from fibronectin resulted in improved cell adhesion to a surface coated with recombinant protein (see Figure 3.2a).[34] Therefore, use of a minimally active short peptide sequence without the context of the flanking amino acids may not recapitulate the native protein interactions and functions.

Multidomain constructs, as enabled by recombinant techniques, add an additional level of control and potential for therapeutic impact. By specifying the number and type of domains, we can create a multifunctional protein and potentially profit from synergistic interactions.[35] One can mix and match a variety of functional and structural domains, as described in Figure 3.1. From this perspective, domains fall into several broad categories, including cell-adhesive, enzyme-reactive, crosslinking, and structural. Domains can also bind soluble factors or matrix components, as discussed above, or even mimic cell-displayed ligands. Depending on the therapeutic application, some or all of these properties may be desirable in a cell-delivery hydrogel, and recombinant technology offers a straightforward means to encoding these functional features in a single material.

Another important consequence of the multidomain design strategy is the interplay between structural domains, which confer mechanical integrity and resilience, and other bioactive domains. The length, flexibility, and stiffness of linking regions between the bioactive domains can alter their biological effect (see Figure 3.2b). Cells are mechanically sensitive to their environment and are hypothesized to sense their surroundings by attempting to deform the matrix, leading to several biochemical changes.[36] Moreover, rearrangement of biologically active ligands is important for many signalling processes,[37] and stiffness and flexibility of structural domains can affect the thermodynamics of cell-induced clustering of bioactive domains. These

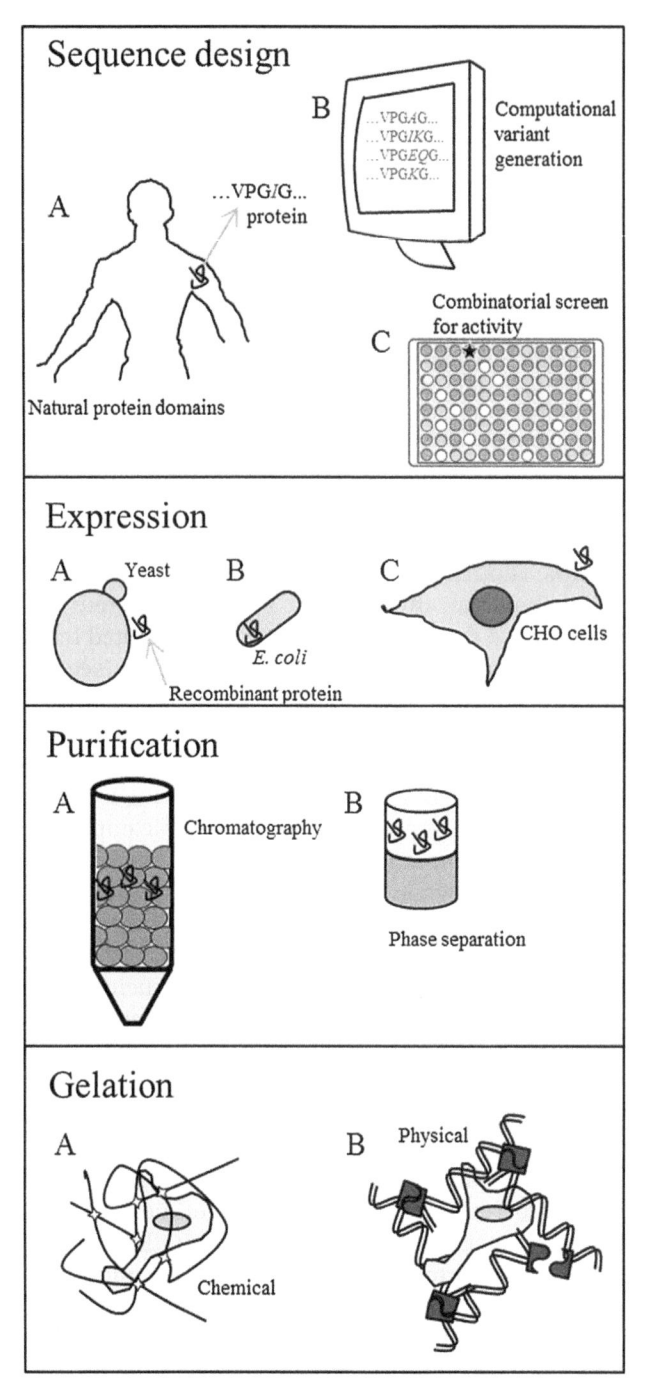

Figure 3.3 Schematic showing steps for producing a recombinant protein hydrogel, from top to bottom. During sequence design, (A) DNA encoding a domain of interest can be cloned from a source organism, *e.g.* elastin

important considerations demand the kind of sequence-level control that is enabled by recombinant materials when optimizing a biomaterial for a specific application.

Multidomain complexity has the potential to be leveraged to enable hydrogels that dynamically adapt to the implant environment in biologically sophisticated ways. While such crosslinks can be created at cell encapsulation, they can also be engineered to form in response to enzymes, making it possible to engineer on-demand deformability.[38] For example, placement of crosslinking or self-assembling domains near a bioactive domain is hypothesized to lead to mechanical deformation of the bioactive sequence,[31] which may dynamically affect bioactive function.[39] Likewise, incorporating enzyme-sensitive domains into a recombinant protein hydrogel allows cell- or physician-demanded changes in network architecture *via* secretion or injection of sequence-specific proteases, which lead to changes in the hydrogel's mechanical and diffusion properties.[27] On the other hand, it is possible for enzyme-specific cleavage domains to be placed on both sides of another domain to enable its liberation in response to secretion of the target enzyme (see Figure 3.2c). This would solubilize the domain and allow its internalization in response to the event driving secretion of the specific enzyme.[40] Many proteins in nature, such as growth factors and integrin ligands, have sharply distinct activities depending on whether they are immobilized or soluble.[41,42] Recombinant proteins enable these complex material responses due to their multidomain capacity.

3.3 From Concept to Protein-Engineered Cell Delivery

As hydrogel components, recombinant proteins enable the material to interact with host and delivered cells in a tailored, biologically sophisticated way, as discussed above. In general, making recombinant proteins involves amino acid sequence selection and optimization, synthesis of a gene that encodes the desired sequence, protein synthesis in an appropriate host organism, and purification of the recombinant protein (see Figure 3.3). In the specific context of hydrogels for cell delivery, once we have pure recombinant

---

from human skin; alternatively or in conjunction, (B) computationally generated sequences, based on knowledge- or physics-based potentials, can be used; likewise, (C) random or derived variants can be combinatorially screened for biological function. Engineered sequences are then expressed in a heterologous host, commonly one of the following: (A) yeast, (B) *E. coli* bacteria, or (C) Chinese hamster ovary (CHO) cells. After expression, the protein is purified; common techniques include (A) chromatography and (B) phase separations. For cell-delivery applications, therapeutic cells can be encapsulated *via* (A) chemical or (B) physical gelation.

proteins, different proteins can be blended for additional or optimized functionality, followed by gelation and cell encapsulation. Clearly, not all sequences of amino acids are suitable for cell encapsulation or even hydrogel formation. The identity of the protein's sequence must be carefully designed not only to provide the desired biological activity but also to ensure useful gelation mechanics. To achieve these two goals, proteins for hydrogels must include bioactive domains to elicit biological activity as well as crosslinking domains and relatively hydrophilic inter-crosslink sequences to control the network structure and hence gel mechanics.

To homogeneously distribute cells throughout a hydrogel for delivery applications, the cells are typically mixed with the polymer components in the sol state before inducing a phase change to the gel state. The sol to gel phase transition can either occur *in situ* (*i.e.* after injection into the body) or *ex situ* (*i.e.* outside the body). Many protein-based hydrogels, such as collagen, gel *in situ*: changing ionic strength, pH, or temperature to that of the implant site drives a sol–gel transition in the material, entrapping co-injected cells. For *ex situ* crosslinking, the cell-loaded material can be either (1) implanted as a preformed cell–gel construct during an open surgery or (2) injected if the hydrogel can quickly shear-thin upon extrusion and quickly re-gel *in vivo*. Both *in situ* and *ex situ* gelation strategies have been successfully used in the design of protein-engineered hydrogels.

There are three protein engineering approaches to designing recombinant amino acid sequences: direct selection of native sequences, combinatorial mutagenesis followed by screening, and computational algorithm-aided design. A variety of domains have evolved in nature that may be useful in the formation of engineered hydrogels, and some of these domains can be directly cut-and-pasted into engineered proteins.[43] This approach is often used as a starting point for protein-biomaterials design, since complete *de novo* specification of an amino acid sequence represents an immense design space that grows exponentially with sequence length. Combinatorial strategies for sequence refinement, which involve screening over a diverse library of randomized mutants, can be used to adjust the affinity of protein interactions.[44] From the perspective of domain-level engineering, sequence refinement of flanking residues is sometimes used to improve the domain's ability to remain stable and functional even when removed from its native context.[45] On the other hand, computational methods use knowledge- or physics-based potentials to infer a sequence most likely to give rise to a desired structure.[46] This can be especially helpful for designing binding domains.[47] It is not uncommon in the field of protein engineering to blend these approaches; for example, computational strategies can be used to prescribe a more narrow range of amino acids to randomize.[48] Using combinations of these strategies, protein-engineered bioactive ligands,[49] self-assembly domains,[50] and enzyme targets[51] have been developed.

After selecting the amino acid sequences of interest, plasmids encoding for the cognate nucleic acid sequences or genes are synthesized. Traditionally assembled by molecular cloning techniques, plasmids that contain genes for recombinant proteins have several common characteristics, including an origin of replication and an additional gene to enable plasmid maintenance. For example, plasmids that are designed for bacterial expression will contain an antibiotic resistance gene, which permits only those bacteria that maintain the plasmid to grow in a medium supplemented with the antibiotic. Contemporary genes are more commonly synthesized by using polymerase chain reactions to merge oligonucleotides and then cloned into commercially available plasmids.[52]

A plasmid embodying a synthetic gene of interest is then transfected into a host cell type chosen for optimal expression and purification. Bacteria, such as *Escherichia coli*, and yeast, such as *Pichia pastoris*, can be fermented in suspension culture, which improves volumetric yield, though mammalian cells are dominant in the pharmaceutical industry.[53] Although higher-order host cells (such as those derived from mammals) frequently demand more complex media, they also have greater capability to fold complex proteins, including those with disulfide bonds and glycosylation. An additional potential advantage of eukaryotic hosts is that the recombinant protein can be designed so that it is secreted into the medium. This simplifies purification of the target protein and avoids cell lysis, which often introduces difficult-to-remove impurities such as endotoxins. In general, impurities can be removed by ionic, reverse phase, and/or affinity chromatography.[54] While chromatography techniques are most common in lab-scale purification processes, highly scalable purification schemes may rely on phase separation protocols. For example, inverse temperature cycling[55] can be used to purify target proteins with lower critical temperature solubility. As another example, high molecular weight polyethylene glycol can be used to drive coacervation of proteins from solution.[56]

After sequence design, gene synthesis, protein expression, and purification (all of which are general steps for making any engineered recombinant protein), engineered proteins for cell-delivery applications must be further processed *via* a cytocompatible gelation mechanism, a function that must be designed into their amino acid sequence. The gelation strategy can rely either on the formation of covalent or non-covalent crosslinking, often termed 'chemical' or 'physical' hydrogels, respectively. Formation of a hydrated, covalent network requires the use of crosslinking chemicals or enzymes. For example, a crosslinking molecule with multiple reactive groups can be used to covalently link several proteins together. Often, this involves a small-molecule crosslinker that targets amino acid moieties such as the primary amine side chain of lysine.[57] A key limitation of this strategy for cell-delivery applications is that the crosslinking reaction must be cytocompatible, since the crosslinking reaction has to take place in the immediate presence of cells.[58] General reactivity, *e.g.*, crosslinking of cell-displayed proteins, and side-products of these chemical reactions can damage encapsulated cells. In the case of enzymatic hydrogels, specific amino acid sequences are targeted for ligation in a manner that is mechanistically similar to natural systems.[59] Enzymes favour specific substrates, which reduces toxic off–target interactions. Another limitation of the covalent gelation strategy is that to make these systems injectable, the kinetics of the crosslinking reaction must be carefully controlled so that the injection process is complete before the final network has formed.

Recombinant proteins can also be engineered to encapsulate cells by non-covalent means such as affinity or self-assembly interactions between proteins.[60] Similar to enzymatic hydrogels, specific sequences enable crosslinking interactions; however, these non-covalent interactions are transient and sensitive to physiological changes in the environment.[61] Through careful engineering of the affinity interactions, these materials can be designed to self-assemble *in situ* in response to the physiological environment.[62] In this strategy, cells are premixed with the engineered protein in the sol phase, and then gelation occurs *in vivo*. As an alternative cell-delivery route, physical hydrogels can be engineered to be injectable by virtue of shear-thinning and self-healing properties.[63] In this strategy, cells are encapsulated in the gel phase *in vitro* prior to injection into the body. For both types of cell encapsulation and delivery, the kinetics of gelation, which are relevant to cell placement within the injection site, can be adjusted by amino acid sequence design.

3.4 Case Studies: Recent Developments and Applications of Recombinant Hydrogels

We further demonstrate these ideas by describing the design features of several recently developed protein-engineered hydrogels that have achieved control over both gelation and bioactivity.

3.4.1 Elastin

A common strategy in the design of protein-engineered hydrogels for cell delivery is to mimic structural components of the native ECM. In many cases, these ECM structural proteins already contain domains that (1) crosslink or self-assemble and (2) hydrate in the physiological environment.[64,65] One protein that attracted early interest is elastin, an essential constituent of many human tissues such as skin, lung, smooth muscle, and vasculature.[66] Elastin confers the required resiliency to these tissues, all of which undergo mechanical strain. In pioneering work, Urry and others recombinantly expressed elastin's elastomeric sequence, which consists of serial repeats of valine–proline–glycine–valine–glycine.[67] In the native protein, the second valine is frequently substituted with one, but sometimes up to three, other amino acids. Later work showed that this serial repeat, known as elastin-like protein (ELP), could be further modified to induce crosslinking and water retention at physiological conditions.[68] While elastin's elastomeric domain has been intensely researched, other domains from the native protein, such as the lysyl oxidase recognition/crosslinking domain, have been less frequent targets of investigation.[69] Recently, full-length native tropoelastin has been recombinantly expressed with various modifications to alter the material properties.[70] These diverse engineering strategies have advanced our understanding of how to use recombinant elastin in cell-delivery applications.

Recent advances in elastin-based hydrogels have enabled cell encapsulation using a variety of approaches. With implantation in mind, these techniques are cytocompatible, and some are apt for form fitting and injection. Chemical crosslinking was originally enabled in ELP's elastomeric domain by replacing the second valine with a lysine.[71] In these biomaterials, the reactive primary amine is targeted for conjugation with an intermediate reactive group that covalently links multiple proteins. Unfortunately, many of these reactions require cytotoxic solvents or produce toxic by-products, making them unsuitable for cell encapsulation. In more recent work, crosslinking agents that are efficient in physiological buffer and cell compatible have been reported.[58] Adjusting crosslinking stoichiometry facilely produces hydrogels with a range of mechanical stiffnesses relevant to human tissue. For example, ELP materials with elastic moduli ranging from fetal to adult heart tissue were designed to study the impact of biomaterial mechanics on the contractility of encapsulated embryonic stem cell-derived cardiomyocytes (see Figure 3.4).[72]

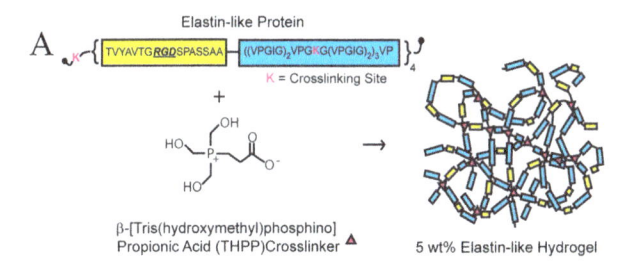

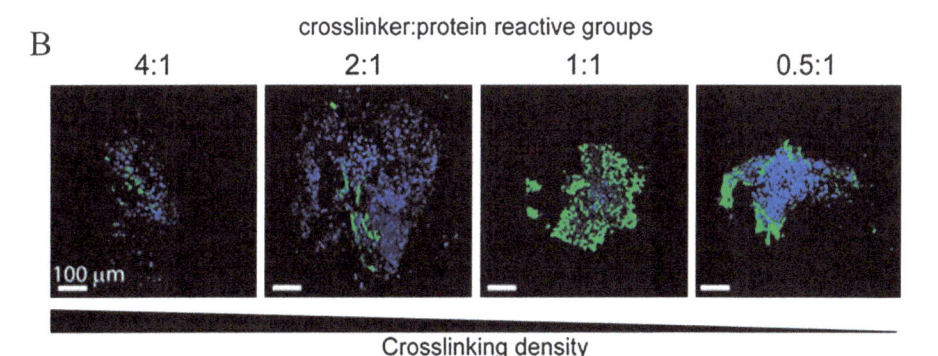

Figure 3.4 (A) Elastin-like proteins (ELPs) have been engineered to enable cytocompatible, amine-reactive encapsulation of human embryonic stem cell embryoid bodies with THPP. (B) Hydrogel crosslinking stoichiometry influences the distribution of cardiac troponin T (green) with respect to nuclear stain (blue) within embryoid bodies. Reproduced from ref. 72 with permission.

Another cytocompatible ELP crosslinking scheme relies on the spontaneous covalent coupling between aldehyde and hydrazide groups at physiological conditions.[73] In this strategy, glutamic acid residues replace a fraction of the second valines in the ELP pentapeptide repeat. After ELP expression, a portion of the protein is grafted with aldehyde groups at the glutamate positions, while another portion of the protein is modified with hydrazide at the glutamate positions. Mixing these two modified proteins together at physiological conditions results in spontaneous gelation without generating by-products. Mechanical stiffness of the resulting hydrogel can be specified by controlling the extent of conjugation to glutamate.

As a general design strategy for ELP-based materials, careful selection of the amino acid replacing the second valine in the elastomeric pentapeptide changes the hydrophobicity of the recombinant protein without significantly changing its elastomeric properties.[74] This selection influences swelling, which is relevant to cell delivery for several reasons.[75] First, swelling is inversely proportional to the concentration of incorporated bioactive domains, which governs binding equilibrium.[76] Second, swollen hydrogels have a greater effective mesh size, which affects diffusion of soluble molecules. On the other hand, ELP hydrophobicity also affects ELP's lower critical solubility temperature (LCST).[77] Increasing the environmental temperature above the LCST causes ELP hydrogels to markedly de-swell, thereby reversibly increasing protein concentration and decreasing mesh size. A number of temperature-sensitive applications in tissue engineering have been developed to exploit this reversible LCST behaviour.[78]

ELPs have been engineered to include not only the elastomeric domains but also the crosslinking domains from native elastin. In this engineered protein, the reactive portion is also a primary amine from lysine, but this lysine is flanked by a mostly polyalanine sequence from native elastin.[79] The chemical difference between the crosslinking and elastomeric domains causes ordered coacervation, placing the crosslinking amino acids in close proximity to each other. This coacervation enables efficient zero-length crosslinking, in the sense that gelation occurs simply by oxidizing reactive lysine.[69] In a related strategy for cell encapsulation, ELP was engineered to be susceptible to crosslinking by the enzyme transglutaminase by inserting truncated glutamine substrates between the elastomeric and lysine-based crosslinking domains.[80]

More recently, analysis of the full-length tropoelastic sequence led to the identification of a bridge domain between the crosslinking and elastomeric domains that facilitates hydrogel formation and biomimetic fibrillogenesis.[81] Manipulating the amino acid sequence of this domain altered the self-assembly of the elastin-based hydrogel, from amorphous to fibrillar. Elastin fibrils are a topographical cue, as seen by increased proliferation of fibroblasts in the fibrillar, *i.e.* native-like, tropoelastin hydrogels.[81] An interesting consequence of coacervation-assisted crosslinking is that these hydrogels feature resiliency after deformation of over 90%,[69] making them especially useful for applications that impose frequent deformations.

Overall, groups studying elastin as a hydrogel constituent have devised several gelation techniques that lend flexibility to potential applications in cell implantation.

Recombinant techniques are leveraged to achieve not only gelation and mechanical control, but also phenotypic control *via* signals from engineered domains. For example, several groups have developed ELPs that include the arginine–glycine–aspartate cell-binding domain.[82] The inclusion of this domain has been shown to promote cell adhesion and proliferation.[83] Blending this bioactive ELP with a non-adhesive ELP variant containing a non-cell-binding arginine–aspartate–glycine scrambled domain, enables tuning of the strength of the cellular response without changing any other material properties, such as stiffness or conductivity. Several other signalling domains have been incorporated into ELPs, including a cell-adhesion domain from laminin.[84] Also of note is the potential bioactivity of ELP's elastomeric domain. When covalently bound, the elastomeric pentapeptide has been reported to promote endothelial production of nitric oxide and smooth-muscle contractility.[85] These phenotypic markers are indicative of organ-appropriate function.

ELP hydrogels have also been modified to enable enzymatic remodelling. Several target sequences sensitive to urokinase and tissue plasminogen activators,[86] enzymes that are secreted during angiogenesis and nerve regeneration,[87,88] were incorporated within ELP sequences to create a family of engineered proteins with differing degradation kinetics. Furthermore, by blending ELPs bearing enzymatic substrates of different sensitivity, gels with complex degradation profiles were created. This was confirmed by the release of entrapped molecules,[86] demonstrating a proof-of-principle for using protein-engineered hydrogels for complex, tissue-responsive drug delivery.

Native tropoelastin's C-terminal heparin-binding domain offers another strategy for controlled release. A variety of growth factors (*e.g.* some isoforms of vascular endothelial growth factor), bind and are presented by heparin.[89] By encapsulating cells with heparin and an ELP variant that contains a heparin-binding domain, hydrogels can be made to retain and potentiate growth factors that bind heparin.[90,91] Similarly, recombinant tropoelastin also has domains for binding other proteins, such as fibrillins, that are rich in cell-active amino acid sequences.[92]

For cell delivery, recombinant, modular ELP hydrogels offer several strategies for achieving desired gelation and bioactive properties. Within current formulations, cells can potentially be entrapped within a hydrogel, maintained by supportive bioactive ligands, and released as the hydrogel's constituent proteins are degraded and resorbed.

3.4.2 Collagen

Like elastin, collagen, specifically types I and III, is an essential structural component of diverse human tissues, providing mechanical strength and stiffness. Native collagen primarily consists of glycine–X–Y repeats, where X and Y can be any amino acid, most commonly proline or hydroxyproline, a modified amino acid. This repetitive sequence contributes to self-assembled triple helices, known as tropocollagen, between three independent collagen peptides.[93] With the aid of fibroblasts, tropocollagen further assembles into hydrated, hierarchical structures.[94] Given the native molecule's capacity for self- and cell-assisted assembly and its well-known bioactivity, collagen appears to be a promising constituent protein for cell-carrying hydrogels.

In a variety of clinical applications, such as cosmetic surgery and vocal cord reconstruction, collagen harvested from animal sources is already used as an implantable biomaterial.[95,96] Bovine collagen is currently harvested from genetically and geographically closed herds,[97] which increases reproducibility and decreases risk of contamination from pathogens. Decellularization techniques are also employed to eliminate the possibility of host reactions to xenogenic cells.[98] Despite the utility of bovine collagen, recombinant collagen offers specific advantages in control of gelation mechanics and bioactivity. Harvested collagens require harsh treatment to transform the mature matrix into a solution phase (sol) for suspending therapeutic cells. After denaturation, collagen sol is maintained at low pH. Collagen hydrogels slowly form as protein is transferred to a buffer with physiological salt and pH, allowing the clinician to encapsulate therapeutic cells and inject the incipient hydrogel. However, due to the complexity of native collagen, some amino acid sequences remain denatured even after regelation, marking the implanted hydrogel as foreign.[99]

Recent developments in recombinant collagen hydrogels focus on engineering self-assembly and specific bioactivity into simplified, collagen-mimetic proteins. In one example, a yeast strain was co-transformed with genes coding for prolyl hydroxylase, which enables synthesis of hydroxyproline—an essential component of native collagen, and for a collagen-like protein (CLP) engineered to self-assemble.[100] This recombinant CLP consists of a hydrophilic domain flanked on both sides by an assembly domain.[101] The hydrophilic domain was based on the stereotypical repeat (glycine–X–Y) of native collagen, but hydrophobic amino acids like alanine were replaced with glutamine or asparagine. The assembling domain consisted of (glycine–hydroxyproline–glycine)$_9$, which spontaneously forms triple helices at physiological pH. The combination of these domains enabled the formation of a canonical hydrogel, which could be further tailored to control the gel erosion rate and the pH of sol–gel transition.[102]

In addition to gelation mechanics, enzyme-mediated degradation and release can be controlled by sequence-level engineering of recombinant CLPs.[100] Sensitivity to general serum proteases was reduced by replacing arginines in the glycine–X–Y repeats. The hydrophilic domain that is exposed in physiological conditions was further engineered to impart selective sensitivity to MMP-2, -9, and -14, proteases that are expressed during wound remodelling.[103]

Another approach to making recombinant CLPs is to express variants of streptococcal collagen-like protein in *E. coli*. Serotype M28 spontaneously assembles into a collagen-like triple helix, but is biologically inert, meaning that it contains no known mammalian cell-binding sites. Such a blank slate is useful for tissue engineering applications, where we want to specify material–cell interactions completely. Currently lacking higher-level assembly, this protein has been conjugated into a poly(ethylene glycol) hydrogel using photoacrylate chemistry,[104] but there is no obvious reason why it could not be incorporated as a functional domain into other self-sufficient recombinant protein hydrogels. To bestow the CLP with cell adhesivity, the cell-adhesive GFOGER peptide sequence was engineered into this otherwise bioinert material. Given the utility of existing collagen hydrogels, we can expect contributions from protein-engineered variants in the future.

3.4.3 Resilin-like Proteins

For some clinical applications, such as replacement of heart valves, vocal cords, or load-bearing cartilage, implantable biomaterials must first and foremost satisfy stringent mechanical requirements, like appropriate stiffness, viscoelasticity, resilience, and fatigue. A relatively new class of recombinant proteins that have been engineered to achieve specified mechanical properties is based on the protein resilin. Resilin is an essential component of the leg and wing joints of insects that is, as natively expressed, a hydrated network.[105] During insect flight, resilin enables roughly 97% of muscle force to be transmitted as lift.[106] Resilin's special role in insect physiology has attracted the interest of materials scientists; as early as 2005, recombinant resilin was expressed in *E. coli* from the corresponding *Drosophila melanogaster* gene.[107] In this pioneering work, resilin was cross-linked into a 20% protein-mass hydrogel *via* ruthenium-catalysed ultraviolet oxidation of tyrosine, which converts individual tyrosine amino acids into dityrosine. Although this engineered material is more resilient than synthetic elastomers, such as polybutadiene, obstacles to its clinical use such as cell compatibility and biodegradability have only been more recently addressed in a family of materials known as resilin-like proteins (RLPs).

Engineered RLPs depart from native resilin in many ways, but these can be summarized in terms of the designers' goals: cytocompatible crosslinking and fully controlled protein bioactivity. Resilient mechanical domains, which consist of elastomeric repeats from native resilin, are commonly derived from Drosophila,[108,109] but have also been derived from *Anopheles gambiae*.[110] Starting from a Drosophila-sourced RLP, Kiick and coworkers replaced tyrosine with lysine in the amino acid sequence, enabling application of known, amine-reactive, and cytocompatible crosslinking chemistries. Cells were successfully encapsulated using amine-reactive small molecules, such as tris(hydroxymethyl)phosphine (THP) and four-armed poly(ethylene glycol) that was end-modified with vinyl sulfone moieties (PEG-VS).[111,112]

64   Chapter 3

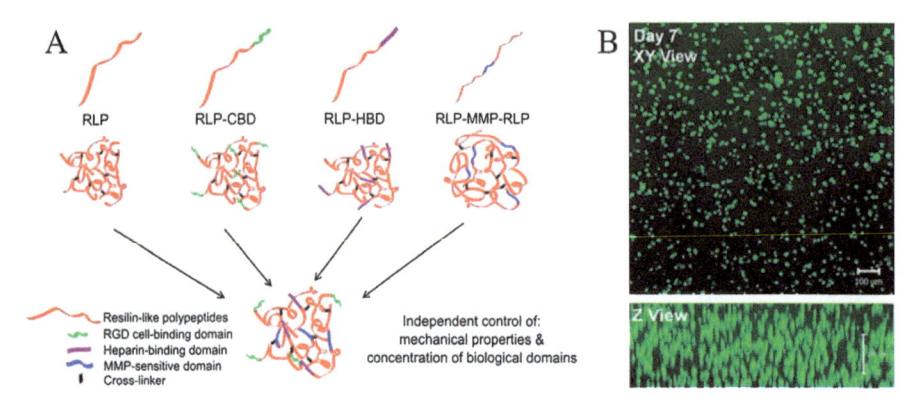

Figure 3.5 (A) Resilin-like proteins (RLPs) have been engineered to support amine-reactive, cytocompatible crosslinking and to allow multifactorial incorporation of diverse biologically active domains simply by blending variant RLP-bioactive fusions. (B) Crosslinking with tris(hydroxymethyl phosphine) (THP) allows for live (indicated by green staining) incorporation of singlet mesenchymal stem cells. Reproduced from ref. 113 with permission.

To control mechanical stiffness, crosslinking stoichiometry and protein content were varied. Despite moving away from native-like tyrosine crosslinking, these RLP hydrogels were over 90% resilient at strains of up to twice the original length. Mesenchymal stem cells were viably encapsulated in these hydrogels.[107]

Controlling protein-mediated bioactivity of RLPs so as to, for example, manipulate proliferation and differentiation of encapsulated stem cells, has been accomplished through domain-level sequence design. Elastomeric domains from native resilin, which are hydrophilic and disordered, were abstracted from the native protein and incorporated as structural elements in RLPs.[112] This may reduce immunogenicity by eliminating fixed-structure epitopes that can be recognized by antibodies. After eliminating undesired activity, bioactive sequences that may lead to therapeutic outcomes were selectively placed in the sequence. Currently, cell-adhesive domains from fibronectin and MMP-sensitive domains from collagen have been incorporated (see Figure 3.5).[113]

3.4.4 Mixing-Induced Two-Component Hydrogels (MITCH)

As an alternative to mimicking the native matrix, recombinant technology can be used to create protein-based hydrogels that are orthogonal to the endogenous cellular microenvironment. From this blank slate, additional bioactive domains can be engineered into the protein. Heilshorn and coworkers demonstrated this approach with mixing-induced two-component hydrogels (MITCH).[60] MITCH consists of two recombinant proteins: one protein has a series of proline-rich domains, which reversibly bind to several

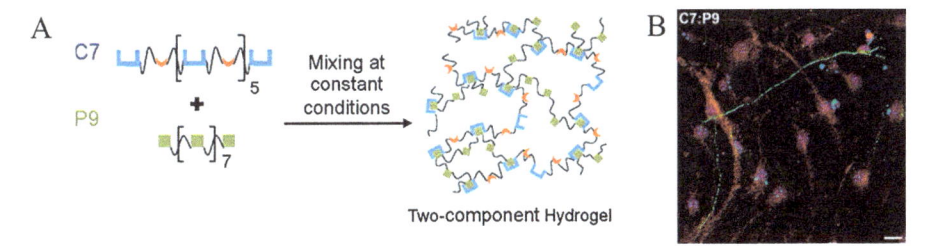

Figure 3.6 (A) Mixing-induced two-component hydrogel (MITCH) is capable of encapsulating cells without chemical reagents or any changes in pH, ionic strength, or temperature, due to its complementary self-assembling hetero-domains: the WW domain (blue) and the proline-rich domain (olive). Cell-adhesive domains (tangerine) are included to promote integrin binding. Reprinted from ref. 115 with permission. Copyright 2011, American Chemical Society. (B) MITCH can encapsulate diverse cell types; here, adult neural stem cells are differentiated for 6 days (red, glial marker GFAP; green, neuronal marker MAP2; yellow, neural progenitor marker nestin; blue, nuclei, DAPI). Reproduced from ref. 30 with permission. Copyright 2009, National Academy of Sciences of the USA.

WW domains contained in the second engineered protein to form physically crosslinked assemblies. Given that these domains are derived from intracellular proteins,[114] interactions with other extracellular proteins are not expected. Hydrophilic peptide spacer regions connect complementary crosslinking domains in order to control the distance between crosslinks, with direct consequence for elastic modulus.

Protein engineering allows precise control of the gelation mechanics of MITCH, a shear-thinning, self-healing hydrogel of mechanical stiffness relevant to soft tissues like marrow, brain, and adipose tissue. These qualities make the material especially apt for injectable cell therapies related to these tissues. Both the gel stiffness and the kinetics of self-healing were found to depend on the strength of binding interactions, which were adjusted by altering the primary amino acid sequence of the physical crosslinking domains to increase binding affinity.[115] The stoichiometric ratio of the two engineered proteins can also be adjusted to change the threshold protein concentration required to form a gel,[115] allowing formation of hydrogels with 5% protein content by mass. The low protein content reduces invasiveness of the treatment by minimizing injection of foreign material.

In terms of bioactivity, a minimal arginine–glycine–aspartate–serine peptide has been incorporated into one of the component proteins. As formulated, MITCH is a cell-maintaining material, as shown with neuronal stem cells, adipose-derived stem cells, and vascular endothelial cells (see Figure 3.6).[60,116] To seed cells within MITCH, only simple mixing of cells and the two components is required. Once encapsulated in MITCH, these cells can be injected through a thin needle without changing their phenotype.

By absorbing the stress from extensional flow in the needle and at its junctions, MITCH may protect therapeutic cell types as it shear-thins.[117] Unlike many other shear-thinning gels, MITCH self-heals in the body without requiring any temperature, pH, or ionic shift. Recombinant engineering of MITCH's complementary binding domains allows this minimally invasive cell-delivery process.

3.5 Challenges and Opportunities in Clinical Translation

Recombinant protein hydrogels, while currently of research interest, have not been used for cell delivery in the clinical setting. As in the use of therapeutic cells itself,[118,119] there are practical obstacles to using recombinant proteins as biomaterials. Work is under way to address these challenges. For example, recombinant proteins are expected to be significantly more expensive to produce than synthetic polymers, despite the pharmaceutical industry's established ability to produce recombinant proteins at yields of grams per litre.[120] Properties of recombinant hydrogels mitigate this disadvantage. First, like other hydrogels, the recombinant hydrogels described above are over 90% water. Second, engineered protein domains have the capacity for molecular recognition. Combined with therapeutic cells, these materials are likely to be highly potent, which may decrease the raw amount needed for therapeutic efficacy. Overall, with regards to bioprocess engineering, recombinant proteins are well ahead of cell therapies in areas such as production scale, cost, sterility, and safety. Further development is required in both fields.

Another challenge arises from the inherent complexity of protein-engineered materials: how can we predictively design and optimize these materials to meet specific therapeutic needs? Engineering amino acid sequences for therapeutic function is a challenge that has attracted many researchers,[121,122] but another issue is how do we combine these bioactive and structural domains into a hydrogel for cell delivery. Given the complex material requirements for cell-delivery applications, approaches to model and optimize multifactorial biomaterials are being developed, which could be readily applied to recombinant protein hydrogels.[123,124] Protein-engineered hydrogels can be thought of as having several biomaterials parameters that influence cells, such as ligand density, mechanical stiffness, mesh size, and proteolytic sensitivity, leading to a combinatorial design space. Building from combinatorial optimization used in the chemical industry, statistical approaches can be used to infer an optimum in a multifactorial design space from a relatively small, carefully scattered set of test conditions.[125] A variety of algorithms exist for statistical optimization.[126,127] Highly defined biomaterials, like recombinant protein hydrogels, have tremendous potential use in cell therapies, but their inherent complexity demands more sophisticated strategies for optimization and application.

Having established the high level of molecular-level design control that recombinant technology enables, we note that recent studies of cell-based therapeutics suggest that a high degree of specificity and control may be required to direct cell and tissue plasticity. On the cellular level, it is likely that cell-carriers will need to feature a greater degree of complexity to effectively guide and/or control therapeutic cells post-implantation, especially for therapies involving the delivery of stem cells. To remain functional, stem cells require a complex array of protein-based signals, broadly including cell-displayed and matrix-displayed ligands as well as mechanical cues.[128,129]

Besides providing a complex niche, recombinant protein hydrogels are especially suited for use as cell-responsive biomaterials. For example, a protein-based cell carrier could stimulate maintenance and amplification of tissue-specific stem cells with one enzyme-labile component and recruit mature matrix with another, more durable component. Broadly speaking, materials that release soluble factors or delivered cells on demand are of great clinical interest.[130] Recombinant proteins can achieve this by combining sequestering domains, cell-adhesion ligands, and enzyme targets within a single specified sequence. Careful design of enzyme targets, which directly affect proteolysis kinetics, and crosslinking domains, which affect diffusion and availability, can help control temporal delivery. Several current challenges in translating stem cell delivery to the clinic may require this kind of temporal control. For example, frequently, implanted stem cells do not persist, let alone integrate into the regenerating tissue due to the hostile wound environment.[131] Temporary delivery of suppressive factors is one strategy being investigated to address this problem.[132] Complex feedback systems that are built into the material are likely to be required to stimulate regenerative pathways, which, in other animals, prominently feature chemical gradients and sequential presentation of multiple factors.[133,134]

3.6 Conclusion

Recent and potential advances in recombinant protein hydrogels make this class of materials an exciting prospect for cell delivery. Using the machinery of life, materials scientists have been able to create fully specified hydrogels with precise gelation mechanics and bioactive functions. Thanks to the genetic control and chemical diversity of recombinant proteins, many cytocompatible gelation strategies have been applied to encapsulate therapeutically relevant cell types. Progress has also been made in engineering some of the biochemically potent protein interactions found in native tissue into the constituent proteins of recombinant hydrogels. Current studies have shown that using full-length recombinant proteins has the potential to bring sophisticated design strategies, including spatiotemporal control of biochemical and mechanical interactions, to bear on delivery of therapeutic cells. Overall, the engineering and application of recombinant protein hydrogels to clinical needs is a fertile field for materials scientists and clinicians alike.

References

1. M. Gnecchi, Z. Zhang, A. Ni and V. J. Dzau, *Circ. Res.,* 2008, **103**, 1204.
2. D. S. Krause, N. D. Theise, M. I. Collector, O. Henegariu, S. Hwang, R. Gardner, S. Neutzel and S. J. Sharkis, *Cell,* 2001, **105**, 369.
3. R. F. Selden, M. J. Skoskiewicz, K. B. Howie, P. S. Russell and H. M. Goodman, *Science,* 1987, **236**, 714.
4. J. Ankrum and J. M. Karp, *Trends Mol. Med.,* 2010, **16**, 203.
5. E. A. Rayment and D. J. Williams, *Stem Cells,* 2010, **28**, 996.
6. D. A. Liberles, S. A. Teichmann, I. Bahar, B. Ugo, J. Bloom, E. Bornberg-Bauer, L. J. Colwell, A. P. de Koning, N. V. Dokholyan and J. Echave, *Protein Sci.,* 2012, **21**, 769.
7. S.-H. Kim, J. Turnbull and S. Guimond, *J. Endocrinol.,* 2011, **209**, 139.
8. M. A. Olayioye, R. M. Neve, H. A. Lane and N. E. Hynes, *EMBO J.,* 2000, **19**, 3159.
9. D. A. Lauffenburger and A. F. Horwitz, *Cell,* 1996, **84**, 359.
10. A. H. Brivanlou and J. E. Darnell Jr, *Science,* 2002, **295**, 813.
11. C. H. Thomas, J. H. Collier, C. S. Sfeir and K. E. Healy, *Proc. Natl. Acad. Sci. U. S. A.,* 2002, **99**, 1972.
12. P. A. DiMilla, J. A. Stone, J. A. Quinn, S. M. Albelda and D. A. Lauffenburger, *J. Cell Biol.,* 1993, **122**, 729.
13. C. Y. Chung, S. Lee, C. Briscoe, C. Ellsworth and R. A. Firtel, *Proc. Natl. Acad. Sci. U. S. A.,* 2000, **97**, 5225.
14. R. Flaumenhaft, D. Moscatelli, O. Saksela and D. B. Rifkin, *J. Cell. Physiol.,* 1989, **140**, 75.
15. L. Macri, D. Silverstein and R. A. F. Clark, *Adv. Drug Delivery Rev.,* 2007, **59**, 1366.
16. J. I. Jones, A. Gockerman, W. H. Busby, C. Camacho-Hubner and D. R. Clemmons, *J. Cell Biol.,* 1993, **121**, 679.
17. R. Flaumenhaft, D. Moscatelli and D. B. Rifkin, *J. Cell Biol.,* 1990, **111**, 1651.
18. A. J. Engler, S. Sen, H. L. Sweeney and D. E. Discher, *Cell,* 2006, **126**, 677.
19. J. W. Weisel, *Biophys. J.,* 1986, **50**, 1079.
20. C. Wu, V. Keivenst, T. E. O'Toole, J. A. McDonald and M. H. Ginsberg, *Cell,* 1995, **83**, 715.
21. X. Liu, Y. Zhao, J. Gao, B. Pawlyk, B. Starcher, J. A. Spencer, H. Yanagisawa, J. Zuo and T. Li, *Nat. Genet.,* 2004, **36**, 178.
22. D. Aeschlimann, O. Kaupp and M. Paulsson, *J. Cell Biol.,* 1995, **129**, 881.
23. Z. Keresztessy, É. Csősz, J. Hársfalvi, K. Csomós, J. Gray, R. N. Lightowlers, J. H. Lakey, Z. Balajthy and L. Fésüs, *Protein Sci.,* 2006, **15**, 2466.
24. B. Steffensen, L. Häkkinen and H. Larjava, *Crit. Rev. Oral Biol. Med.,* 2001, **12**, 373.
25. P. Van Lint and C. Libert, *J. Leukocyte Biol.,* 2007, **82**, 1375.
26. H. A. Klok, *Macromolecules,* 2009, **42**, 7990.
27. C. Lin and K. S. Anseth, *Proc. Natl. Acad. Sci. U. S. A.,* 2011, **108**, 6380.

28. E. A. Phelps, N. Landázuri, P. M. Thulé, W. R. Taylor and A. J. García, *Proc. Natl. Acad. Sci. U. S. A.,* 2010, **107**, 3323.
29. S. G. Peisajovich, J. E. Garbarino, P. Wei and W. A. Lim, *Science,* 2010, **328**, 368.
30. Z. Rao, P. Handford, M. Mayhew, V. Knott, G. G. Brownlee and D. Stuart, *Cell,* 1995, **82**, 131.
31. S. C. Heilshorn, J. C. Liu and D. A. Tirrell, *Biomacromolecules,* 2005, **6**, 318.
32. A. P. Silverman, A. M. Levin, J. L. Lahti and J. R. Cochran, *J. Mol. Biol.,* 2009, **385**, 1064.
33. J. D. Pédelacq, S. Cabantous, T. Tran, T. C. Terwilliger and G. S. Waldo, *Nat. Biotechnol.,* 2005, **24**, 79.
34. E. Fong and D. A. Tirrell, *Adv. Mater.,* 2010, **22**, 5271.
35. D. Schubert, *Trends Cell Biol.,* 1992, **2**, 63.
36. N. Wang, J. D. Tytell and D. E. Ingber, *Nat. Rev. Mol. Cell Biol.,* 2009, **10**, 75.
37. S. De, O. Razorenova, N. P. McCabe, T. O'Toole, J. Qin and T. V. Byzova, *Proc. Natl. Acad. Sci. U. S. A.,* 2005, **102**, 7589.
38. S. Kim and K. E. Healy, *Biomacromolecules,* 2003, **4**, 1214.
39. E. Klotzsch, M. L. Smith, K. E. Kubow, S. Muntwyler, W. C. Little, F. Beyeler, D. Gourdon, B. J. Nelson and V. Vogel, *Proc. Natl. Acad. Sci. U. S. A.,* 2009, **106**, 18267.
40. C. N. Salinas and K. S. Anseth, *Biomaterials,* 2008, **29**, 2370.
41. Y. H. Shen, M. S. Shoichet and M. Radisic, *Acta Biomater.,* 2008, **4**, 477.
42. D. G. Stupack and D. A. Cheresh, *J. Cell Sci.,* 2002, **115**, 3729.
43. C. Khosla and P. B. Harbury, *Nature,* 2001, **409**, 247.
44. W. S. Sandberg and T. C. Terwilliger, *Proc. Natl. Acad. Sci. U. S. A.,* 1993, **90**, 8367.
45. R. H. Kimura, Z. Cheng, S. S. Gambhir and J. R. Cochran, *Cancer Res.,* 2009, **69**, 2435.
46. S. M. Lippow and B. Tidor, *Curr. Opin. Biotechnol.,* 2007, **18**, 305.
47. G. Grigoryan, A. W. Reinke and A. E. Keating, *Nature,* 2009, **458**, 859.
48. C. A. Voigt, S. L. Mayo, F. H. Arnold and Z. G. Wang, *Proc. Natl. Acad. Sci. U. S. A.,* 2001, **98**, 3778.
49. S. C. Heilshorn, K. A. DiZio, E. R. Welsh and D. A. Tirrell, *Biomaterials,* 2003, **24**, 4245.
50. E. R. Ballister, A. H. Lai, R. N. Zuckermann, Y. Cheng and J. D. Mougous, *Proc. Natl. Acad. Sci. U. S. A.,* 2008, **105**, 3733.
51. M. K. McHale, L. A. Setton and A. Chilkoti, *Tissue Eng.,* 2005, **11**, 1768.
52. D. M. Hoover and J. Lubkowski, *Nucleic Acids Res.,* 2002, **30**, e43.
53. F. M. Wurm, *Nat. Biotechnol.,* 2004, **22**, 1393.
54. J. Nilsson, S. Ståhl, J. Lundeberg, M. Uhlén and P. Nygren, *Protein Expression Purif.,* 1997, **11**, 1.
55. D. E. Meyer and A. Chilkoti, *Nat. Biotechnol.,* 1999, **17**, 1112.
56. J. M. Harris, *Poly(Ethylene Glycol) Chemistry: Biotechnical and Biomedical Applications,* Springer, New York, 1992.
57. R. A. McMillan and V. P. Conticello, *Macromolecules,* 2000, **33**, 4809.

58. C. Chung, K. J. Lampe and S. C. Heilshorn, *Biomacromolecules*, 2012, **13**, 3912.
59. N. E. Davis, S. Ding, R. E. Forster, D. M. Pinkas and A. E. Barron, *Biomaterials*, 2010, **31**, 7288.
60. C. T. S. Wong Po Foo, J. S. Lee, W. Mulyasasmita, A. Parisi-Amon and S. C. Heilshorn, *Proc. Natl. Acad. Sci. U. S. A.*, 2009, **106**, 22067.
61. G. Schreiber, *Curr. Opin. Struct. Biol.*, 2002, **12**, 41.
62. N. Q. Tran, Y. K. Joung, E. Lih, K. M. Park and K. D. Park, *Biomacromolecules*, 2010, **11**, 617.
63. L. Haines-Butterick, K. Rajagopal, M. Branco, D. Salick, R. Rughani, M. Pilarz, M. S. Lamm, D. J. Pochan and J. P. Schneider, *Proc. Natl. Acad. Sci. U. S. A.*, 2007, **104**, 7791.
64. D. Bedell-Hogan, P. Trackman, W. Abrams, J. Rosenbloom and H. Kagan, *J. Biol. Chem.*, 1993, **268**, 10345.
65. F. H. Silver, J. W. Freeman and G. P. Seehra, *J. Biomech.*, 2003, **36**, 1529.
66. L. Debelle and A. Tamburro, *Int. J. Biochem. Cell Biol.*, 1999, **31**, 261.
67. D. T. McPherson, C. Morrow, D. S. Minehan, J. Wu, E. Hunter and D. W. Urry, *Biotechnol. Prog.*, 1992, **8**, 347.
68. P. J. Nowatzki and D. A. Tirrell, *Biomaterials*, 2004, **25**, 1261.
69. F. W. Keeley, C. M. Bellingham and K. A. Woodhouse, *Philos. Trans. R. Soc. London, Ser. B*, 2002, **357**, 185.
70. B. Vrhovski, S. Jensen and A. S. Weiss, *Eur. J. Biochem.*, 1997, **250**, 92.
71. E. R. Welsh and D. A. Tirrell, *Biomacromolecules*, 2000, **1**, 23.
72. C. Chung, E. Anderson, R. R. Pera, B. L. Pruitt and S. C. Heilshorn, *Soft Matter*, 2012, **8**, 10141.
73. U. M. Krishna, A. W. Martinez, J. M. Caves and E. L. Chaikof, *Acta Biomater.*, 2012, **8**, 988.
74. D. W. Urry, T. Hugel, M. Seitz, H. E. Gaub, L. Sheiba, J. Dea, J. Xu and T. Parker, *Philos. Trans. R. Soc. London, Ser. B*, 2002, **357**, 169.
75. N. A. Peppas, J. Z. Hilt, A. Khademhosseini and R. Langer, *Adv. Mater.*, 2006, **18**, 1345.
76. D. A. Lauffenburger and J. Linderman, *Receptors: Models for Binding, Trafficking, and Signaling*, Oxford University Press, New York, 1996.
77. A. Ribeiro, F. J. Arias, J. Reguera, M. Alonso and J. C. Rodríguez-Cabello, *Biophys. J.*, 2009, **97**, 312.
78. L. Klouda and A. G. Mikos, *Eur. J. Pharm. Biopharm.*, 2008, **68**, 34.
79. C. M. Bellingham, M. A. Lillie, J. M. Gosline, G. M. Wright, B. C. Starcher, A. J. Bailey, K. A. Woodhouse and F. W. Keeley, *Biopolymers*, 2003, **70**, 445.
80. S. Bozzini, L. Giuliano, L. Altomare, P. Petrini, A. Bandiera, M. T. Conconi, S. Farè and M. C. Tanzi, *J. Mater. Sci.: Mater. Med.*, 2011, **22**, 2641.
81. G. C. Yeo, C. Baldock, A. Tuukkanen, M. Roessle, L. B. Dyksterhuis, S. G. Wise, J. Matthews, S. M. Mithieux and A. S. Weiss, *Proc. Natl. Acad. Sci. U. S. A.*, 2012, **109**, 2878.
82. A. Panitch, T. Yamaoka, M. J. Fournier, T. L. Mason and D. A. Tirrell, *Macromolecules*, 1999, **32**, 1701.

83. K. S. Straley and S. C. Heilshorn, *Soft Matter,* 2009, **5**, 114.
84. K. S. Straley and S. C. Heilshorn, *Front. Neuroeng.,* 2009, **2**, 9.
85. P. H. Blit, W. G. McClung, J. L. Brash, K. A. Woodhouse and J. P. Santerre, *Biomaterials,* 2011, **32**, 5790.
86. K. S. Straley and S. C. Heilshorn, *Adv. Mater.,* 2009, **21**, 4148.
87. P. Mignatti and D. B. Rifkin, *Enzyme Protein,* 1996, **49**, 117.
88. L. B. Siconolfi and N. W. Seeds, *J. Neurosci.,* 2001, **21**, 4336.
89. W. H. Burgess and T. Maciag, *Annu. Rev. Biochem.,* 1989, **58**, 575.
90. Y. Tu, S. M. Mithieux, N. Annabi, E. A. Boughton and A. S. Weiss, *J. Biomed. Mater. Res., Part A,* 2010, **95**, 1215.
91. T. J. Broekelmann, B. A. Kozel, H. Ishibashi, C. C. Werneck, F. W. Keeley, L. Zhang and R. P. Mecham, *J. Biol. Chem.,* 2005, **280**, 40939.
92. P. Booms, A. Ney, F. Barthel, G. Moroy, D. Counsell, C. Gille, G. Guo, R. Pregla, S. Mundlos and A. J. P. Alix, *J. Mol. Cell. Cardiol.,* 2006, **40**, 234.
93. M. J. Buehler, *Curr. Appl. Phys.,* 2008, **8**, 440.
94. F. H. Silver, J. W. Freeman and G. P. Seehra, *J. Biomech.,* 2003, **36**, 1529.
95. C. N. Ford, P. A. Staskowski and D. M. Bless, *Laryngoscope,* 1995, **105**, 944.
96. T. S. Alster and T. B. West, *Plast. Reconstr. Surg.,* 2000, **105**, 2515.
97. S. L. Matarasso, *Plast. Reconstr. Surg.,* 2007, **120**, 17S.
98. E. Rieder, G. Seebacher, M. T. Kasimir, E. Eichmair, B. Winter, B. Dekan, E. Wolner, P. Simon and G. Weigel, *Circulation,* 2005, **111**, 2792.
99. D. Michaeli and H. H. Fudenberg, *Clin. Immunol. Immunopathol.,* 1974, **2**, 153.
100. M. W. T. Werten, T. J. van den Bosch, R. D. Wind, H. Mooibroek and F. A. de Wolf, *Yeast,* 1999, **15**, 1087.
101. H. Teles, T. Vermonden, G. Eggink, W. Hennink and F. de Wolf, *J. Controlled Release,* 2010, **147**, 298.
102. W. Shen, K. Zhang, J. A. Kornfield and D. A. Tirrell, *Nat. Mater.,* 2006, **5**, 153.
103. T. H. Vu and Z. Werb, *Genes Dev.,* 2000, **14**, 2123.
104. E. Cosgriff-Hernandez, M. Hahn, B. Russell, T. Wilems, D. Munoz-Pinto, M. Browning, J. Rivera and M. Höök, *Acta Biomater.,* 2010, **6**, 3969.
105. M. Burrows and G. P. Sutton, *J. Exp. Biol.,* 2012, **215**, 3501.
106. T. Weis-Fogh, *J. Exp. Biol.,* 1973, **59**, 169.
107. C. M. Elvin, A. G. Carr, M. G. Huson, J. M. Maxwell, R. D. Pearson, T. Vuocolo, N. E. Liyou, D. C. C. Wong, D. J. Merritt and N. E. Dixon, *Nature,* 2005, **437**, 999.
108. G. Qin, S. Lapidot, K. Numata, X. Hu, S. Meirovitch, M. Dekel, I. Podoler, O. Shoseyov and D. L. Kaplan, *Biomacromolecules,* 2009, **10**, 3227.
109. M. B. Charati, J. L. Ifkovits, J. A. Burdick, J. G. Linhardt and K. L. Kiick, *Soft Matter,* 2009, **5**, 3412.
110. J. N. Renner, Y. Kim, K. M. Cherry and J. C. Liu, *Protein Expression Purif.,* 2012, **82**, 90.
111. C. L. McGann, E. A. Levenson and K. L. Kiick, *Macromolecules,* 2013, **214**, 203.

112. L. Li, S. Teller, R. J. Clifton, X. Jia and K. L. Kiick, *Biomacromolecules,* 2011, **12**, 2302.
113. L. Li, Z. Tong, X. Jia and K. L. Kiick, *Soft Matter,* 2013, **9**, 665.
114. M. J. Macias, S. Wiesner and M. Sudol, *FEBS Lett.,* 2002, **513**, 30.
115. W. Mulyasasmita, J. S. Lee and S. C. Heilshorn, *Biomacromolecules,* 2011, **12**, 3406.
116. A. Parisi-Amon, W. Mulyasasmita, C. Chung and S. C. Heilshorn, *Adv. Healthcare Mater.,* 2013, **2**, 428.
117. B. A. Aguado, W. Mulyasasmita, J. Su, K. J. Lampe and S. C. Heilshorn, *Tissue Eng., Part A,* 2011, **18**, 806.
118. K. C. Wollert and H. Drexler, *Nat. Rev. Cardiol.,* 2010, **7**, 204.
119. L. Naldini, *Nat. Rev. Genet.,* 2011, **12**, 301.
120. F. Schmidt, *Appl. Microbiol. Biotechnol.,* 2004, **65**, 363.
121. D. M. Fowler, C. L. Araya, S. J. Fleishman, E. H. Kellogg, J. J. Stephany, D. Baker and S. Fields, *Nat. Methods,* 2010, **7**, 741.
122. S. Lutz, *Curr. Opin. Biotechnol.,* 2010, **21**, 734.
123. J. R. Smith, A. Seyda, N. Weber, D. Knight, S. Abramson and J. Kohn, *Macromol. Rapid Commun.,* 2004, **25**, 127.
124. Y. Mei, K. Saha, S. R. Bogatyrev, J. Yang, A. L. Hook, Z. I. Kalcioglu, S. W. Cho, M. Mitalipova, N. Pyzocha and F. Rojas, *Nat. Mater.,* 2010, **9**, 768.
125. J. P. Jung, J. V. Moyano and J. H. Collier, *Integr. Biol.,* 2011, **3**, 185.
126. L. Harmon, *J. Mater. Sci.,* 2003, **38**, 4479.
127. J. C. Meredith, A. P. Smith, A. Karim and E. J. Amis, *Macromolecules,* 2000, **33**, 9747.
128. K. Saha, J. F. Pollock, D. V. Schaffer and K. E. Healy, *Curr. Opin. Chem. Biol.,* 2007, **11**, 381.
129. D. E. Discher, D. J. Mooney and P. W. Zandstra, *Science,* 2009, **324**, 1673.
130. R. Langer, *Science,* 1990, **249**, 1527.
131. V. F. M. Segers and R. T. Lee, *Nature,* 2008, **451**, 937.
132. K. Krick, M. Tammia, R. Martin, A. Höke and H. Q. Mao, *Curr. Opin. Biotechnol.,* 2011, **22**, 741.
133. B. S. Ding, D. J. Nolan, P. Guo, A. O. Babazadeh, Z. Cao, Z. Rosenwaks, R. G. Crystal, M. Simons, T. N. Sato and S. Worgall, *Cell,* 2011, **147**, 539.
134. E. M. De Robertis, *Sci. Signaling,* 2010, **3**, pe21.

CHAPTER 4

The Instructive Role of Biomaterials in Cell-Based Therapy and Tissue Engineering

ROANNE R. JONES[†], IAN W. HAMLEY, AND CHE J. CONNON*

School of Chemistry, Food and Pharmacy, University of Reading, Reading RG6 6AD, UK
*E-mail: r.jones@qmul.ac.uk; I.w.hamley@reading.ac.uk; c.j.connon@reading.ac.uk

4.1 Introduction to Cell-Based Therapies

Cell-based therapy (CBT) is a disruptive platform technology that encompasses a variety of approaches for the exogenous delivery of cells (including but not limited to stem cells) *via* transplantation. CBT has great potential when the innate or intrinsic ability of organs fails in the repair and regeneration process needed to combat disease and has the unique ability of restoring life-supporting or essential functions in the human body. CBT is thus a major focus in the field of translational regenerative medicine and differs from its counterpart 'tissue engineering' in that the cells are not typically constrained within a tissue construct formed *ex vivo* prior to transplantation; that said, the differences between the two can become blurred or that CBT can include tissue engineering if the purpose of the construct is to be transplanted.

Stem cells in the human body represent an endogenous system of cell repair and regeneration, and so with the ever-evolving advances in science

---

[†]Currently at Barts Cancer Institute, Queen Mary, University of London, London EC1M 6BQ, UK.

RSC Soft Matter No. 2
Hydrogels in Cell-Based Therapies
Edited by Che J. Connon and Ian W. Hamley
© The Royal Society of Chemistry 2014
Published by the Royal Society of Chemistry, www.rsc.org

and stem cell biology it is logical that such cells are used for their primary function in therapies.[1] Indeed, bone-marrow-derived stem cells have been used clinically for over 40 years in the regeneration of various cell types, demonstrating their potential for use in other debilitating diseases and potential for CBT.

The term 'stem cell' originates from the late 19th century and its first appearance in scientific literature was in 1868; since then varying propositions and definitions have been formulated but a clear-cut definition is now well-understood:

- Self-renewal: stem cells have the capacity to proliferate almost indefinitely.[2]
- Differentiation: stem cells have the ability to differentiate into a range of specialized cells, such as blood or neural cells, from their undifferentiated state.

The stem cell source is usually characteristic of the potency to differentiate into specific or multiple cell types. The tissue type from which stem cells are derived leads to the differentiation of a specific set of cells.[3] Stem cells can be derived both from embryonic cell sources such as the inner cell mass of *in vitro* fertilized embryos, and from adult cell sources including a range of tissues and organs in the adult body. Embryonic stem cells are pluripotent; they have the potential to differentiate into any cell type of the three germ layers.[3] In comparison, adult stem cells are multipotent and differentiate into a more limited range of cell types;[1] their major role involves the repair of tissues and organs. Consequently the more specific types of adult stem cells are related to their tissue origin: for example, haematopoietic stem cells (HSCs) can differentiate into any type of blood cell, and mesenchymal stem cells (MSCs) have the ability to differentiate into a range of cells including bone, cartilage and fat cells.

A relatively novel type of stem cells, derived from the direct reprogramming of somatic cells in 2007, are induced pluripotent stem cells (iPSCs).[4] These cells have the potential to differentiate into any type of human tissue with none of the ethical concerns associated with the use of human embryos, and thus hold great promise for the future of CBT.

There remains some confusion regarding the differences between stem and progenitor cells and the terms are often interchanged; however, it is known that progenitor cells exhibit restricted differentiation potential and limited replication ability.[5] Consequently, it is important to note that the term 'stem cell' is broad and encompasses many different types of cells with differing characteristics. Furthermore, the delivery of adult stem cells in transplantation can be based on autologous (self-donated) or allogeneic cells. Patient-specific therapies are autologous techniques, whereas allogeneic cells are those from another human donor. Each of these cell types has contrasting advantages and disadvantages, which are exploited in the application of CBT in human disease.

4.1.1 Potential Clinical Applications and Pharmaceutical Industry Involvement

Given the vast array of tissues serving as stem cell sources,[3] there is huge potential for many CBTs to be used in regenerative medicine to manage, treat and potentially cure numerous diseases. The current CBT pipeline encompasses development programmes for a range of human diseases.[6] These include cardiovascular diseases such as myocardial infarction and stroke; blood disorders such as sickle cell anaemia; musculoskeletal diseases; neural diseases such as spinal cord injury, Parkinson's disease and motor neuron disease; endocrine diseases including diabetes as well as retinal degenerative diseases and other ocular injuries.[3] The need for clinical and commercial success in this area of research is apparent given the high expectations of patients suffering from such diseases, for which there is currently high clinical unmet need.[7]

The provision of CBT should be a major focus for pharmaceutical companies[8] given the niche market for this emerging sector. However, the lengthy and somewhat risky process associated with the translation of such therapies to the marketplace often hinders the involvement of companies whereby return on investment is not clear-cut.[8]

Human clinical trials are an essential and a fundamental step in the drug discovery and development process for any potential CBT. Phase I and II clinical trials are important in assessing the safety of a treatment in patients, ensuring no harm is caused, while phases III and IV determine the efficacy of a treatment; if successful, these trials lead on to the mass production of such therapies before making them available to patients. The first phase of clinical trials poses the greatest ethical and human protection challenges, which is a particular issue for novel therapies that have not previously been tested in humans. Finally following the lengthy clinical trial process, and also the demonstration of good manufacturing practice (GMP), a government licence is required before a therapy can be used in patients. It is not uncommon for potential CBT to fail before reaching this stage of development and only approximately 10% (in the case of stem cell therapies) have been successful so far.[9]

Of critical consideration in the development and emergence of CBT is the method of delivery. This decision is based upon the therapeutic condition to be treated as the need for cell regeneration at a particular site usually reflects the site of administration. The controlled delivery of stem cells to specific tissue regions is complex[5] and requires highly trained healthcare professionals to ensure optimum safety and minimum harm to patients[8] as well as the use of advanced technologies to guarantee the desired therapeutic effect is ultimately produced.[5] In cardiovascular diseases, for example, a range of methods have been used including intramyocardial injections, intravenous and intracoronary infusions, as well as transendocardial catheter delivery.[10] The design of delivery vehicles such as scaffolds or tissue matrices is largely determined by the stem cell type to be administered.[11] Scaffolding materials such as biodegradable nanofibres are an option for the delivery of stem cells in a controlled release manner, particularly for conditions requiring bone regeneration; this method also provides physical support to the bone site in such cases.[12] However, hydrogels offer huge potential due to their tractable nature, defined levels of compliance and range of possible constituent materials (both synthetic and natural). Clearly advancements in a range of technologies are crucial to the successful adoption of such cell delivery methods in order to fulfil the current expectations of stem cells in the future of medicine.

4.1.2 Stem Cell Banking

Increasingly, reliable cryopreservation methods and stem cell banks allow the storage of cells for many years thus generating new and exciting opportunities for the progression of CBT and the future of regenerative medicine. Such strategies help to overcome issues such as the limited number of stem cell donors, which is common with respect to allogeneic therapy.[13] It is vital that cryopreservation protocols are continuously improved to ensure the maintenance of stem cell quality and prevention of cell death.[14] However, recently hydrogels have also made a significant contribution to the storage of cells. Efficient transport of cells without affecting their survival and function is a key factor in any practical CBT. The current approach using liquid nitrogen for the transfer of cells requires a short delivery time window, and is technically challenging and financially expensive. A recent study has shown that a variety of cell types, once encapsulated in a hydrogel (alginate gel), can maintain their viability and phenotype for up to 1 week at room temperature in a sealed tube.[15] This has been demonstrated by sending half a million mouse embryonic stem cells *via* surface post using nothing more than a paper envelope and a second-class stamp.

4.2 Biomaterials

Biomaterials, natural or synthetic, are frequently used in the biomedical and pharmaceutical industry and are routinely applied as biological substrates, devices for cell delivery, or dressings for tissue regeneration. They may consist of cues or motifs from biologically active sequences or may be derived in their entirety from natural sources. However, due to the biological origin of these materials there is the potential for variability. Naturally occurring biomaterials such as human amniotic membrane (HAM), for example, often have a limited range of mechanical properties and require a degree of optimization before use.[16] Therefore the requirement to develop and improve materials derived from naturally occurring sources is ever increasing. Research into the engineering of biomaterials is moving towards the use of multidisciplinary approaches at the interface of synthetic and biological chemistry, paving the way for the development of improved substrates for the regeneration of functional tissue structures. This chapter focuses on the application of biological and synthetic biomaterials and the mechanical processes influencing cell functionality.

4.2.1 Natural Biomaterials

Natural biomaterials are templates or structures derived from biological components such as proteins, polysaccharides or components of the extracellular matrix (ECM) surrounding cells or tissues. Essentially, the naturally occurring components in biological systems provide the platforms or recognition sites for cell interaction, and are advantageous as substrates for tissue engineering applications because of their outstanding biocompatibility properties. Equally, the effect of cell–cell interaction is vitally important and can be facilitated by the spatial or temporal control of cells as part of tissue homeostasis *in vivo* or *via* external stimulus *in vitro*. The stimulus provided by biomaterials or substrates may promote the alignment and orientation of cells, facilitating the secretion of chemical cues that allow cells to suitably remodel and regulate their microenvironment. Biomaterials may also contain bioactive ligands or sequences that facilitate cell adhesion and together with material properties can be applied to the expansion or regeneration of cells to produce engineered tissue constructs.

Examples of naturally occurring biomaterials include collagen, fibrin, silk, hyaluronic acid, and chitosan, all of which have previously been applied in corneal, neural, bone, and cartilage-based systems.[17-20] HAM is another example of a naturally occurring biomaterial and has historically been applied as a biological substrate to promote wound healing in therapeutic applications.[21,22] The benefits of HAM have largely been recognized due to the precedent of its use as an ocular surface bandage, and this lends support for the use of this substrate as a material for tissue regeneration.[23-25] The structure of HAM is such that it provides good optical transparency, is non-immunogenic and has antimicrobial and antiviral properties. In addition, it is sufficiently durable for use in surgical applications, maintaining enough flexibility to lie flat on the ocular surface.[26] The biochemical properties of HAM, including the presence of cytokines, growth factors, basement membrane proteins and protease inhibitors,[27-29] are thought to contribute to the successful regeneration of damaged tissue by the promotion of re-epithelialization and the reduction of inflammation, scarring and vascularization.[30-35]

The therapeutic effects of HAM have previously been demonstrated by its application as grafts or substrates for cell delivery to promote re-epithelialization and regeneration of the ocular surface.[30,35-40] However, the structural characteristics of HAM may contribute to its variable clinical outcome[16] and only limited approaches have been taken to improve its mechanical properties.[16,41] The need to develop materials with biomimetic tuneable physical properties for specific therapeutic applications is taking centre stage in tissue engineering and CBT.

The design of biological materials may take inspiration from other naturally occurring functional biological polymers such as collagen, silk or fibrin. Collagen is a major component of the ECM and also one of the most abundant structural proteins within mammalian connective tissue. Collagen fibres are inextensible and therefore provide support and mechanical stability to biological structures while maintaining flexibility.[42] In tendons, collagen fibres occur in parallel aligned arrays; in bone, the fibres are organized in concentric layers which maximize resistance to torsional and compressive stresses.[43,44] The ubiquitous occurrence of collagen throughout the body and its biological diversity of function are a result of the many genetically distinct types of collagen. Collagens may exist in two forms, fibrillar and non-fibrillar, and consist of triple-helical domains which assemble to form long, rod-like structures.[43] These structures are comprised of three separate α chains which each contain a characteristic sequence of amino acids. The structure of the triple helix is stabilized by hydrogen bonding of the amino acid α chains.[45] In fibrillar collagens the triple-helical domain makes up over 95% of the molecule, but in the non-fibril forming collagens the triple-helical domain comprises only a fraction of the molecule's mass.[44] Types I, II, III, V and XI are fibrillar collagens commonly distributed in fibrous stromal matrices.

Type I collagen, made up of α1 and α2 chains, is abundantly distributed throughout connective framework of tissues such as skin, bone and the cornea. The well-ordered collagen fibres within the cornea consists predominantly of types I and V fibrillar collagens.[46] The arrangement of collagen fibrils within this region is such that they form highly aligned tightly packed sheets of ECM, facilitating the transmission of light. Type I collagen has formed the basis of bioinspired polymerized matrices which have previously been demonstrated as appropriate biomimetic structures for the formation of functional tissue constructs.[47-50] As a substrate, collagen is highly biocompatible, with low levels of immunogenicity,[51] making it an excellent structure for tissue engineering applications. The fibrils within polymerized collagen structures (gels) can be compacted by the application of weight to form compressed collagen scaffolds with improved collagen density and mechanical properties.[47,48] The dehydration and fibril compaction that occurs throughout this process, known as plastic compression,[47] facilitates the generation of biphasic structures made up of two interpenetrating phases, an entangled fibrillar network and continuous interstitial solution.[52]

The limitations of naturally occurring biomaterials, *e.g.* HAM, can be found in their restricted range of mechanical properties. The development of biological-inspired biomimetic materials is attracting a great deal of interest. Bioinspired materials like compressed collagen gels may not only be manipulated for specific applications (corneal therapy),[49,53] but may also be applied for transferrable applications in alternative tissue types.[41]

4.2.2 Synthetic Materials in Tissue Engineering Applications

At the interface of synthetic and biomaterials chemistry is the use of synthetic materials for the development of biologically inspired scaffolds for tissue engineering applications. It is an area that is attracting intense research activity. Although these materials are termed synthetic they often take biologically inspired cues from ECM proteins or biologically active peptide sequences.

Synthetic materials offer many benefits for use in tissue engineering applications. They are often clinically approved, consist of mechanically stable structures and can therefore be easily modulated to produce materials with variable structural and compositional properties. However, in order to obtain a biological function it is necessary for these materials to be replaced or substituted by a cell-synthesized native substrate as part of the tissue remodelling process.[54] As such, the requirements of synthetic materials are no longer confined to promoting the growth and maintenance of cells. Tissue engineering is moving towards the development of smart materials with tuneable physical and chemical properties with the potential to elicit controlled cell responses or the direction of stem cell lineage.

The design of synthetic biomaterials to produce artificial tissue constructs for CBT requires an understanding of tissue function in relation to the cell microenvironment. This often requires a multidisciplinary approach, particularly where the chemical or physical properties of materials can be manipulated and applied as cues to stimulate changes in cell behaviour.

Typical synthetic biomaterials include poly(ethylene glycol) (PEG)-based gels, electrospun poly(lactic acid-*co*-glycolic acid) (PLGA), or those containing bioactive ligands such as sugars or integrin binding sites.[55-57] Synthetic materials may be chemically modified in order to incorporate bioactive ligands such as cell adhesion motifs, or to stimulate the assembly of appropriate structures such as nanotapes for cell expansion or encapsulation. Self-assembling materials, for example, are often favourable as they can be used to develop novel cell culture systems. The incorporation of peptide epitopes in these materials provides the capacity for high bioactive signalling presentation with the potential to modulate the topographical or mechanical features or spatial organization of 2D or 3D structures. These sophisticated materials can potentially be regulated and modified for specific tissue types for a number of tissue engineering applications.

Synthetically based amphiphiles are an example of self-assembling nanostructures containing biologically active motifs, typically in the form of peptide residues. As such they are termed peptide amphiphiles.[58,59] The self-assembling properties of peptide amphiphiles are attributed to the presentation of hydrophobic tails (*e.g.* C_{16} chains) and hydrophilic head groups in the form of peptide epitopes in each molecule. Alternatively, peptide amphiphiles may contain hydrophilic tails, such as the bioactive RGDS sequence,[60] and hydrophobic units (*e.g.* Fmoc [*N*-(fluorenyl-9-methoxycarbonyl)]); see Figure 4.1. The epitopes contained within these amphiphilic structures typically consist of biologically derived motifs or sequences, rendering them biologically functional. Self-assembly typically occurs through β-sheet formation in the amino sequence together with hydrophobic stacking to form cylindrical nanostructures.[61] The self-assembly of these nanostructures, comprising nanofibres, facilitates the presentation of hydrophilic peptide epitopes on the fibril surface for cell interaction.

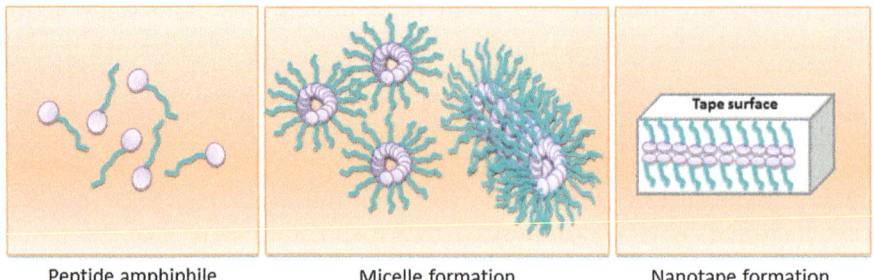

Figure 4.1 Schematic showing the self-assembly of peptide amphiphiles. The assembly of supramolecular scaffolds is attributed to the self-assembling nature of peptide amphiphiles. The formation of nanofibres or nanotapes occurs through the induction of hydrogen bonding together with the collapse of the hydrophobic region of each molecule. The self-assembly of dynamic micelles may also provide structures with a multivalent display of biofunctional motifs for the interaction of cells and tissues.

The use of peptide amphiphiles as self-assembling tuneable scaffolds has previously been demonstrated[61,62] and their use in bone, cartilage, bladder regeneration, and stem cell differentiation has been explored.[61,63–65] The supramolecular structures produced by self-assembling peptide amphiphiles render them good candidates for use as tissue scaffolds or bioactive functional materials. The order and conformation of their structures have the potential to be manipulated for specific cell function, *e.g.* to stimulate the secretion of cell matrix proteins such as collagen. The hydrophobic components of peptide amphiphiles have included hydrocarbon chains, alkyl chains or aromatic Fmoc groups.[66] The self-assembly of peptide amphiphiles containing aromatic Fmoc subunits has been demonstrated[60,67] and their use in cell culture applications has also been shown.[68–70]

The integration of biofunctional cues with synthetic materials typically involves functional peptide groups. Peptide sequences present in fibronectin, collagen and laminin,[71,72] such as RGD, have been used to introduce biofunctionality to materials in a number of systems.[73–76] The interaction of cells with ECM ligand density is known to influence the functionality of cells, which may include migration and adhesion (see Table 4.1). Particularly in the cornea, fibronectin is an intrinsic component of the basement membrane, important for cell adhesion and wound healing,[77,78] and the incorporation of functional ligands to substrates may be applied for the development of artificial corneal tissues.[79]

4.3 Biomaterials in Tissue Engineering

Cell–cell interactions are important for the exchange of vital molecular information to maintain growth, viability and intracellular structural reorganization. However, cells can also obtain a vast amount of information from

Table 4.1 Incorporation of biofunctional cues in synthetic materials to produce bioactive substrates for tissue engineering applications.[a]

Biofunctional structure	Application
RGD	Promotion of dermal fibroblast adhesion[69]
RGDS	Enhanced proliferation and viability of bone-marrow-derived stem cells[80], improved adhesion of bladder cells[63]
KTTKS	Stimulation of collagen secretion in dermal and corneal fibroblasts[81]
Hyaluronic acid	Migration of corneal epithelial cells[17]
Galactose	Improved growth and adhesion of hepatocytes[55]
TGF binding sequences	Cartilage regeneration[64]
PDLLA/bioglass	Proliferation of annulus fibrosus cells and secretion of glycosaminoglycans, collagens I and II[82]

[a] Abbreviations: KTTKS, Lys–Thr–Thr–Lys–Ser; PDLLA, poly-D-L-lactide; RGD, Arg–Gly–Asp; RGDS, Arg–Gly–Asp–Ser; TGF, transforming growth factor.

their microenvironment and surrounding tissue, which serves to regulate or maintain their behaviour *in vivo*. Therefore, the requirements of biomaterials for tissue engineering applications demand many characteristics capable of influencing cell adhesion, migration and differentiation. By incorporating these properties it is possible to produce materials with instructive cues for bioengineering applications (see Figure 4.2). Furthermore, the stimulatory characteristics of these materials may be improved by use in self-assembling nanoparticles, mechanical tractability, bioactive ligand presentation, or topographical modulation.

Due to the regenerative potential of stem cells, the design of materials with instructive cues to direct stem cell fate is the basis for current tissue engineering research. Intensive research in this area has involved the use of substrate mechanical cues and work is now moving towards understanding the mechanisms involved in cell mechanosensing. Understanding the behaviour of cells and their response to environmental cues is equally important for scaffold design as research shifts towards the use of sophisticated or smart materials for tissue regeneration.

4.3.1 Cell Response to Substrate Elasticity

The biochemical and topographical stimulation of substrates has previously been shown to stimulate changes in cell function.[68,69,83–87] However, work has recently shown that substrate-stimulated mechanical stimulation can also be applied to influence cell behaviour. The microenvironment surrounding cells is important for the maintenance of tissue homeostasis, particularly in the growth, migration and lineage specification of stem cells (see Figure 4.2). Work by Engler and colleagues[88,89] has previously demonstrated that the mechanical stimulation elicited by the ECM with which cells are in contact can provide instructive cues for the differentiation of MSCs into alternative

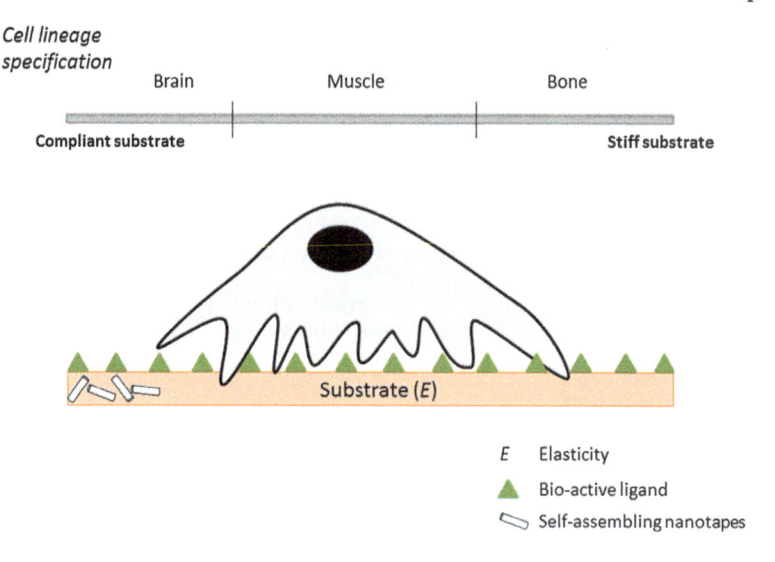

Figure 4.2 Substrate elasticity and cell differentiation. The physical properties of materials can be manipulated to incorporate instructive characteristics. The self-assembling properties of materials can also be used to stimulate the formation of substrates with mechanical tractability. Subsequently the elasticity of materials can be used to influence the differentiation of cells into specific tissue types, while the presentation of biofunctional cues may stimulate cell migration and adhesion. Adapted with permission from ref. 72. Copyright Elsevier 2006.

cell types.[71,73] In these studies, synthetic substrates with a range of elasticity (stiffness) were applied to investigate the differentiation potentiality of MSCs. The development of cell lineages such as bone, lung and heart as a result of substrate stiffness illustrated the sensitivity of cells to the mechanical cues of synthetic biomaterials.[88]

Work has also shown that the mechanical properties of biomaterials can stimulate changes in cell function such as adherence, migration, proliferation and differentiation.[53,90–94] There is also evidence to suggest that substrate stiffness can not only regulate cell phenotype *in vitro* but may also contribute to the physiological maintenance of phenotype as part of the physiological environment of tissues *in vivo*.[95]

These characteristics of biomaterials may be used to manipulate instructive properties of tissue scaffolds, providing them with tuneable physical properties for the regulation of cell behaviour or manipulation of lineage specification.[88,90,91,96] The use of substrate elasticity may be applied to address the limitations of current tissue engineering techniques, particularly where the maintenance of stem cells is required for therapeutic applications, *e.g.* treatment for limbal stem cell deficiency in the cornea. The mechanical regulation of biomaterials may therefore be useful in the design of tuneable scaffolds for tissue-specific cell-based therapies. As stated, current therapeutic applications for stem cell therapies necessitate improvement. Not only are substrates required to be durable and biodegradable, but also they are required to maintain a defined population of cells. The durability and subsequent mechanical integrity of biomaterials are physical characteristics that can be utilized to stimulate the functionality as well as the orientation and spacing of cells in order to regulate an environment tailored for cell survival. In order for substrates to stimulate a response in cell behaviour, an exchange of physical and biochemical information is required which occurs through a number of intra- and extracellular feedback processes. This is often referred to as mechanotransduction.

4.3.2 Structuring of ECM Mimics

As well as mimicking the mechanical properties of the ECM, several important structural features have been shown to be important. These include the presentation of adhesion motifs and the need to incorporate a fibrillar scaffold in place of the native cytoskeleton,[97,98] or reconstructing cytoskeletal components using natural or biomimetic materials. A successful artificial ECM also needs to allow for the incorporation of growth factors, either as soluble factors or by tethering to the matrix. This involves sulfated proteoglycans (glycosaminoglycans) such as heparin or hyaluronic acid which can bind and sequester growth factors such as TGF-β.[97,99–102]

The RGD peptide and variants such as GRGD or RGDS from fibronectin, which binds integrins at the cell surface, has been widely employed in synthetic biomaterials as a minimal cell adhesion motif. It has been demonstrated that a critical spacing of 50–70 nm of adhesion motif clusters is required. Early work showed that an RGD spacing of 440 nm is sufficient for integrin $\alpha_V\beta_3$-mediated fibroblast spreading, but 40 nm is required for focal contact and stress fibre formation.[74] More recent work has explored a more precise control of extracellular ligand presentation, using arrays of gold nanoparticles as anchor points for integrin receptors.[103] This group investigated this further, using a fibronectin-coated AFM cantilever to detach adhered cells from patterned gold nanoparticle surfaces.[104] Ligands were presented with spacings of 28, 50, 90 and 103 nm. For spacings greater than 90 nm, focal contact formation was inhibited and adhesion forces were greatly reduced compared to spacings of 50 nm or less. In recent work, it has been shown, using microstructured patterned gold nanoparticle arrays, that local ligand density has a greater effect than global ligand density on cell adhesion.[105] Cell motility as well as cell adhesion depends on the density of RGD clustering.[106] In addition to control of RGD density at the surface, the influence of RGD spacing in 3D systems such as hydrogels has been examined (see, for example, recent work on PEG-acrylate hydrogels with variable PEG spacer length).[107] The total RGD content is also of course critical,[108] as is its presentation, (*i.e.* the nature of the spacer to the polymer backbone[57]) and the nature of the RGD peptide (linear *vs.* cyclic, multi-arm, *etc.*).[109] The release of RGD *via* enzymatic cleavage, *via* incorporation of matrix metalloproteinase

(MMP) substrates for example, also improves stem cell differentiation.[110] In the field of bulk polymers for tissue engineering, a thorough review of RGD modified polymers is available.[111]

The topography of the substrate at the micro and nano scale can influence cell morphology, adhesion, alignment and proliferation as well as processes of mechanotransduction discussed in the following section.[112,113] The dynamic control of topographical features for instance using pH- or temperature-responsive polymers, or materials responsive to specific chemical triggers, has been the focus of much recent interest.[114–116] This is inspired by efforts to capture aspects of the dynamic remodelling processes of the ECM.

4.3.3 Mechanosensitivity

The physical properties and mechanical forces of the ECM are important for the regulation of tissue homeostasis.[95] Mechanotransduction involves the transformation of substrate-stimulated mechanical cues into intracellular biochemical signalling, which enables cells to sense and respond to external forces and physical constraints in their extracellular environment.[117,118] Part of this process involves the remodelling of the cytoskeleton through contractile proteins and the activation of specific cellular pathways.[92,119] The subsequent changes in cell shape stimulated by substrate stiffness have been shown to have a significant impact on cell spreading and migration.[89,92] It is said that the process by which cells sense substrate-stimulated mechanical forces occurs in three stages: the primary sensory process, the transduction process and the mechanoresponsive pathways that integrate the biochemical signal derived from transduction events;[118] see Figure 4.3.[118]

A network of complex intracellular processes, such as those involving stress fibre formations, the phosphatidylinositol 3 (PI3) kinase pathway,

Mechanosensing	Mechanotransduction	Mechanoresponse
Applied forces from the extracellular matrix	Changes in cytoskeletal organisation (e.g., actin microfilaments)	Transduction geometry dependent interactions to biochemical signals
Composition of the extracellular matrix (e.g., elasticity)	Force-induced changes in the conformation of contractile proteins	Activation of signalling pathways (e.g., HIPPO, PI3 kinase and Rho GTPase activity)
	Changes in cell shape, force generation (e.g., NMMII)	Changes in cell response: differentiation, migration, proliferation

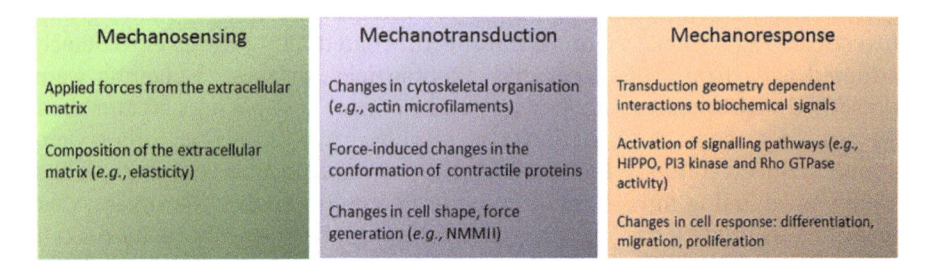

Figure 4.3 Mechanosensing, mechanotransduction and mechanoresponse are three important steps that are involved when cells respond to their environment. Local signals, received from applied forces or the composition of the extracellular matrix, are transduced into biochemical signals eliciting a mechanoresponse from the activation of signalling pathways. The force-induced changes in the conformation of proteins such as contractile proteins or actin microfilaments are converted into cues which regulate the cellular mechanoresponse.

The Instructive Role of Biomaterials in Cell-Based Therapy and Tissue Engineering 85

translocation of the Yes-associated protein (YAP), and the upregulation of focal adhesion complexes such as non-muscle myosin II (NMMII), are said to be involved in the transduction of substrate-stimulated physical and mechanical cues;[88,120,121] (see Figure 4.4). The activation of transcellular adhesion sites causes the binding of actin filaments which initiates contractility of the actomyosin structure. This in turn contributes to the activation of force-stimulated signals, generated from the substrate or microenvironment. These signals are then transduced in a number of intracellular processes in order to stimulate a mechanical response from the cell. The processes involved in the anchorage of cells, pulling and probing the elasticity of their environment, and the interplay of biochemical cues and signalling pathways that ultimately lead to the regulation of cell function and phenotype remain only partially understood.

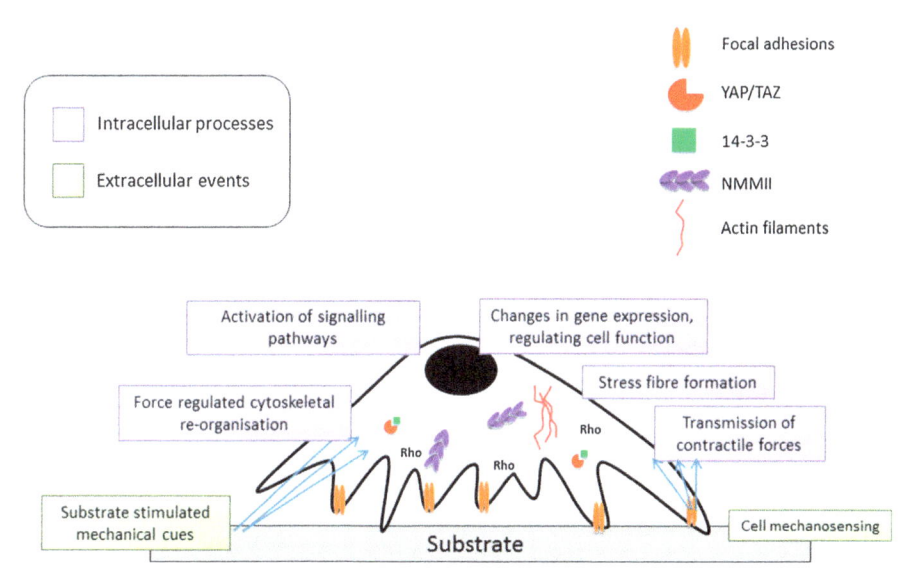

Figure 4.4 Intracellular and extracellular mechanosensing responses. This diagram shows the interplay between biochemical and physical processes that occur during cell mechanotransduction. The transmission of contractile forces through the external mechanical stimulation from substrate or extracellular environment triggers a series of intracellular events involving a complex interaction of proteins and pathways. Cytoskeletal remodelling occurs through the reorganization of actin filaments responsive to the upregulation of focal adhesion proteins which serving as a structural link between the cell cytoskeleton and extracellular matrix. These events may also stimulate changes in cell shape, spreading and migration and subsequent activation of signalling pathways may initiate the reprogramming of cell fate, influencing functional processes such as cell phenotype. Abbreviations: NMMII, non-muscle myosin II; TAZ, transcriptional co-activator with PDZ-binding motif; YAP, Yes-associated protein. Adapted with permission from ref. 81. Copyright Macmillan Publishers Ltd 2006.

4.3.4 Proteins and Pathways

NMMII is an actin-binding protein essential for processes that modulate cell shape and movement.[122] It is comprised of three pairs of peptides in the form of light and heavy chains, functioning to regulate enzyme activity and actin filament movement though the use of adenosine triphosphate (ATP); see Figure 4.5. In addition, the crosslinking and contractile properties of this protein are essential for the formation of focal adhesions and intracellular stress fibres.[123,124]

During mechanotransduction, intracellular sites of adhesion known as focal adhesions are formed which constitute a structural link between the actin cytoskeleton and the ECM. Focal adhesions therefore function as regions of signal transduction, facilitating intracellular communication with the external microenvironment. Cells sense and feel their matrix as a result of

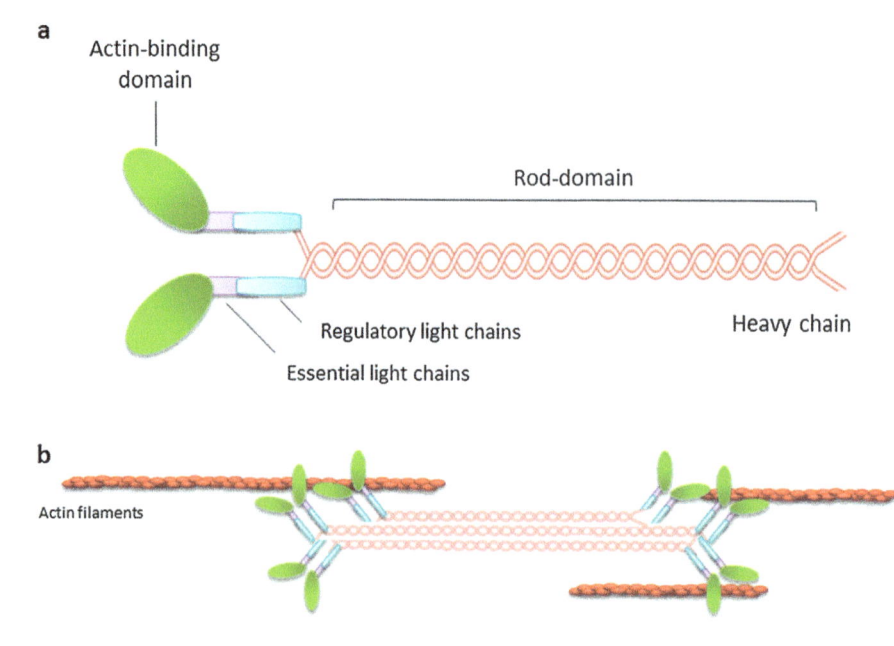

Figure 4.5 Structure and function of non-muscle myosin II (NMMII). (a) The subunit and domain structure of NMMII consists of light and heavy protein chains and globular head domain. The actin-binding ATPase regions of NMMII form part of the globular head which is attached to regulatory and essential light chains. (b) The interaction of rod domains from each NMMII molecule facilitates the assembly of bipolar filaments and subsequent phosphorylation by Rho kinase causes the activation and presentation of ATPase in the head domain of the bipolar filaments. The conformational changes that occur facilitate the binding of actin where the bipolar myosin filaments link the actin filaments together to form thick bundles of cellular structures such as stress fibres. Adapted with permission from ref. 86. Copyright Macmillan Publishers Ltd 2009.

applied forces exerted by the external environment.[118] The activation of these adhesion sites causes the transmission of contractile forces through the cell cytoplasm *via* the crosslinking and contraction of actin filaments.[123] Subsequently, the external forces encountered as a result of cell–substrate communication stimulate changes to the internal architecture of cells through cytoskeleton reorganization, focal adhesion dynamics and signalling transduction processes.[125]

Actin filaments are important for both cell adhesion and cell migration. They are components of the cell cytoplasm and are distributed throughout the cell, constantly becoming polymerized or depolymerized and self-assembling in the presence of crosslinker proteins such as NMMII. Self-assembled actomyosin stress fibres are the major mediators of cell contraction[126] and studies have demonstrated the regulatory mechanisms underlying stress fibre formation. Specifically, Rho-associated protein kinase (ROCK) has been implicated as part of the complex network of signalling that controls the contractility of actomyosin, stress fibre formation and cell differentiation.[119,126] ROCK is a downstream Rho effector, which functions to mediate Rho signalling and actin cytoskeleton reorganization through phosphorylation of several substrates, such as NMMII, that contribute to the assembly of actin filaments and contractility. The Rho family of GTPases are pivotal regulators of several aspects of cell behaviour such as motility, proliferation and apoptosis. As part of these processes they function to regulate the assembly of actin bundles, stress fibres and tensile actomyosin that mediate the adherence of cells to substrates[121] and activate the myosin-dependant contractile forces that are required by cells in substrate sensing[88] which contribute to changes in differentiation-mediated gene expression or lineage specification.[119]

YAP is an intracellular protein that functions as a transcriptional activator and repressor, often coupled with the homologue TAZ (transcriptional co-activator with PDZ-binding motif). YAP is a critical downstream regulatory target in the Hippo signalling pathway (named from the protein kinase Hippo) and has recently been implicated as a sensor and mediator of substrate mechanical cues along with TAZ. The Hippo signalling pathway is involved in the regulation of cell proliferation, apoptosis and organ size in response to changes in cell density. At low cell density, YAP binds to transcription factors to stimulate the transcription of genes that favour cell growth and proliferation. However, as part of cell mechano-sensitivity, YAP is thought to be independent of the Hippo signalling cascade.[121] The regulation of mechanical signals exerted by substrate elasticity requires Rho GTPase activity as well as tension of the actomyosin cytoskeleton.

There is constant interplay between signalling pathways, and the interaction between the Hippo and TGF-β signalling pathways; both involve phosphorylation of the YAP/TAZ transcriptional modulators.[127] Studies have shown that TAZ functions as a transcriptional modulator of mesenchymal stem cell differentiation through 14-3-3 protein binding.[128]

Phosphorylation of YAP/TAZ causes the exposure of binding sites for 14-3-3 and the retention of YAP/TAZ in the cytoplasm of cells. However, upon interaction with 14-3-3 YAP/TAZ are exported to the nucleus.[129-131] The function of YAP/TAZ in cell mechanotransduction suggests that cytoplasmic retention is characteristic of cells responsive to mechanical stimuli on soft substrates.[121]

4.3.5 Relevance to Regenerative Medicine

The ECM provides important cues for the growth and differentiation of stem cells. Particularly, the influence of mechanical cues from the cell microenvironment has large implications for tissue engineering and the development of materials for use in regenerative medicine. Although the process of feedback cues and mechanosensitivity is yet to be understood, the ongoing work in this area highlights the important considerations that should be taken in material design whether it be in 2D or 3D microenvironments. The elasticity of materials or substrates can significantly influence (directly or indirectly) the proliferation, migration and differentiation of cells. Therefore the design of tissue mimics should pay consideration to the forces that cells may feel and sense in response to the stiffness of their microenvironment. In addition this provides the opportunity for multidisciplinary approaches for the development of fine-tuneable cell substrates capable of modulating cell behaviour according to the requirements of therapies, not only by influencing cell phenotype but also by regulating proliferation and migration. The development of biological mimics for therapeutic applications such as tissue regeneration or cell delivery may facilitate positive treatment outcomes if cells are regulated to function comparable to their native tissue. This may also provide the opportunity to introduce ligands and motifs for chemical cues combining biochemical influences with mechanical forces resulting from material elasticity, moving towards smart materials for regenerative therapies. For example, tuneable substrates may be used to modulate cell secretion and matrix deposition, using the substrate material as a template for cells to produce their own microenvironment in a controlled manner.

The effects of substrate mechanical forces on cell phenotype on multipotent stem cell populations such as MSCs have been extensively demonstrated and may be applied in unipotent stem cell populations for the delivery of tissue-specific stem cells in therapeutic applications. An understanding of tissue homeostasis is something that should also be considered because the mechanical environment that cells are exposed to *in vivo* is known to provide regulatory cues for cell behaviour. The compelling evidence demonstrated by previous groups demonstrating the influence of topographical, mechanical and biochemical cues on the differentiation status of cells provides the basis for work investigating the influence of modulative biomaterials for the design and engineering of artificial tissue equivalents.

References

1. S. Bajada, I. Mazakova, J. B. Richardson and N. Ashammakhi, *J. Tissue Eng. Regener. Med.*, 2008, **2**, 169–183.
2. M. Ramalho-Santos and H. Willenbring, *Cell Stem Cell*, 2007, **1**, 35–38.
3. V. Volarevic, B. Ljujic, P. Stojkovic, A. Lukic, N. Arsenijevic and M. Stojkovic, *Br. Med. Bull.*, 2011, **99**, 155–168.
4. K. Takahashi, K. Tanabe, M. Ohnuki, M. Narita, T. Ichisaka, K. Tomoda and S. Yamanaka, *Cell*, 2007, **131**, 861–872.
5. L. Ricotti and A. Menciassi, *IEEE Trans. Biomed. Eng.*, 2013, **60**, 727–734.
6. Datamonitor, 2011, HC00072-002.
7. F.-M. Chen, Y.-M. Zhao, Y. Jin and S. Shi, *Biotechnol. Adv.*, 2012, **30**, 658–672.
8. A. Trounson, N. D. DeWitt and E. G. Feigal, *Stem Cells Transl. Med.*, 2012, **1**, 9–14.
9. EuroStemCell, Taking stem cell therapies to patients: the role of commercialisation, http://www.eurostemcell.org/factsheet/taking-stem-cell-therapies-patients-role-commercialisation, accessed 29/3/13.
10. I. H. Schulman and J. M. Hare, *Regener. Med.*, 2012, **7**, 17–24.
11. N. A. Kouris, J. M. Squirrell, J. P. Jung, C. A. Pehlke, T. Hacker, K. W. Eliceiri and B. M. Ogle, *Regener. Med.*, 2011, **6**, 569–582.
12. Z. Zhang, J. Hu and P. X. Ma, *Adv. Drug Delivery Rev.*, 2012, **64**, 1129–1141.
13. S. M. Broder, R. S. Ponsaran and A. J. Goldenberg, *Transfusion*, 2013, **53**, 679–687.
14. D. Balci and A. Can, *Curr. Stem Cell Res. Ther.*, 2013, **8**, 60–72.
15. B. Chen, B. Wright, R. Sahoo and C. J. Connon, *Tissue Eng., Part C*, 2013, **19**, 568–576.
16. B. Chen, R. R. Jones, S. Mi, J. Foster, S. G. Alcock, I. W. Hamley and C. J. Connon, *Soft Matter*, 2012, **8**, 8379–8387.
17. J. A. Gomes, R. Amankwah, A. Powell-Richards and H. S. Dua, *Br. J. Ophthalmol.*, 2004, **88**, 821–825.
18. S. M. Willerth, K. J. Arendas, D. I. Gottlieb and S. E. Sakiyama-Elbert, *Biomaterials*, 2006, **27**, 5990–6003.
19. Y. Wang, U.-J. Kim, D. J. Blasioli, H.-J. Kim and D. L. Kaplan, *Biomaterials*, 2005, **26**, 7082–7094.
20. M. Gravel, T. Gross, R. Vago and M. Tabrizian, *Biomaterials*, 2006, **27**, 1899–1906.
21. G. Colocho, W. P. Graham, 3rd, A. E. Greene, D. W. Matheson and D. Lynch, *Arch. Surg.*, 1974, **109**, 370–373.
22. I. Mermet, N. Pottier, J. M. Sainthillier, C. Malugani, S. Cairey-Remonnay, S. Maddens, D. Riethmuller, P. Tiberghien, P. Humbert and F. Aubin, *Wound Repair Regen.*, 2007, **15**, 459–464.
23. A. de Rötth, *Arch. Ophthalmol.*, 1940, **23**, 522–525.
24. A. Sorsby, J. Haythorne and H. Reed, *Br. J. Ophthalmol.*, 1947, **31**, 409–418.

25. A. Sorsby and H. M. Symons, *Br. J. Ophthalmol.*, 1946, **30**, 337–345.
26. C. J. Connon, J. Doutch, B. Chen, A. Hopkinson, J. S. Mehta, T. Nakamura, S. Kinoshita and K. M. Meek, *Br. J. Ophthalmol.*, 2010, **94**, 1057–1061.
27. M. F. Champliaud, G. P. Lunstrum, P. Rousselle, T. Nishiyama, D. R. Keene and R. E. Burgeson, *J. Cell Biol.*, 1996, **132**, 1189–1198.
28. N. Koizumi, T. Inatomi, C. J. Sotozono, N. J. Fullwood, A. J. Quantock and S. Kinoshita, *Curr. Eye Res.*, 2000, **20**, 173–177.
29. B. K. Na, J. H. Hwang, J. C. Kim, E. J. Shin, J. S. Kim, J. M. Jeong and C. Y. Song, *Placenta*, 1999, **20**(Supplement 1), 453–466.
30. J. Shimazaki, H. Y. Yang and K. Tsubota, *Ophthalmology*, 1997, **104**, 2068–2076.
31. A. Azuara-Blanco, C. T. Pillai and H. S. Dua, *Br. J. Ophthalmol.*, 1999, **83**, 399–402.
32. S. H. Lee and S. C. Tseng, *Am. J. Ophthalmol.*, 1997, **123**, 303–312.
33. M. Kubo, Y. Sonoda, R. Muramatsu and M. Usui, *Invest. Ophthalmol. Visual Sci.*, 2001, **42**, 1539–1546.
34. H. M. Woo, M. S. Kim, O. K. Kweon, D. Y. Kim, T. C. Nam and J. H. Kim, *Br. J. Ophthalmol.*, 2001, **85**, 345–349.
35. D. F. Anderson, P. Ellies, R. T. F. Pires and S. C. G. Tseng, *Br. J. Ophthalmol.*, 2001, **85**, 567–575.
36. N. Koizumi, T. Inatomi, T. Suzuki, C. Sotozono and S. Kinoshita, *Arch. Ophthalmol.*, 2001, **119**, 298–300.
37. R. J. Tsai, L. M. Li and J. K. Chen, *N. Engl. J. Med.*, 2000, **343**, 86–93.
38. I. R. Schwab, M. Reyes and R. R. Isseroff, *Cornea*, 2000, **19**, 421–426.
39. N. Koizumi, T. Inatomi, T. Suzuki, C. Sotozono and S. Kinoshita, *Ophthalmology*, 2001, **108**, 1569–1574.
40. S. C. G. Tseng, P. Prabhasawat, K. Barton, T. Gray and D. Meller, *Arch. Ophthalmol.*, 1998, **116**, 431–441.
41. S. Mi, A. L. David, B. Chowdhury, R. R. Jones, I. W. Hamley, A. M. Squires and C. J. Connon, *Tissue Eng., Part A*, 2012, **18**, 373–381.
42. P. Bornstein and H. Sage, *Annu. Rev. Biochem.*, 1980, **49**, 957–1003.
43. A. J. Bailey, R. G. Paul and L. Knott, *Mech. Ageing Dev.*, 1998, **106**, 1–56.
44. E. D. Hay, *Cell Biology of Extracellular Matrix*, Plenum Press, New York, 1991.
45. W. Traub and K. A. Piez, *Adv. Protein Chem.*, 1971, **25**, 243–352.
46. D. A. Newsome, J. Gross and J. R. Hassell, *Invest. Ophthalmol. Visual Sci.*, 1982, **22**, 376–381.
47. R. A. Brown, M. Wiseman, C. B. Chuo, U. Cheema and S. N. Nazhat, *Adv. Funct. Mater.*, 2005, **15**, 1762–1770.
48. S. Mi, B. Chen, B. Wright and C. J. Connon, *J. Biomed. Mater. Res., Part A*, 2010, **95**, 447–453.
49. S. Mi, V. V. Khutoryanskiy, R. R. Jones, X. Zhu, I. W. Hamley and C. J. Connon, *J. Biomed. Mater. Res., Part A*, 2011, **99**, 1–8.
50. S. Mi, B. Chen, B. Wright and C. J. Connon, *Tissue Eng., Part A*, 2010, **16**, 2091–2100.

51. E. Bell, B. Ivarsson and C. Merrill, *Proc. Natl. Acad. Sci. U. S. A.*, 1979, **76**, 1274–1278.
52. V. H. Barocas and R. T. Tranquillo, *J. Biomech. Eng.*, 1997, **119**, 137–145.
53. R. R. Jones, I. W. Hamley and C. J. Connon, *Stem Cell Res.*, 2012, **8**, 403–409.
54. R. A. Brown, in *Future Strategies for Tissue and Organ Replacement*, ed. J. M. Polak, L. L. Hench and P. Kemp, Imperial College Press, London, 2002, pp. 51–78.
55. R. F. Ambury, C. L. R. Merry and R. V. Ulijn, *J. Mater. Chem.*, 2011, **21**, 2901–2908.
56. K. J. Aviss, J. E. Gough and S. Downes, *Eur. Cells Mater.*, 2010, **19**, 193–204.
57. C. N. Salinas and K. S. Anseth, *J. Tissue Eng. Regener. Med.*, 2008, **2**, 296–304.
58. I. W. Hamley, *Soft Matter*, 2011, **7**, 4122–4138.
59. J. B. Matson and S. I. Stupp, *Chem. Commun.*, 2012, **48**, 26–33.
60. V. Castelletto, C. M. Moulton, G. Cheng, I. W. Hamley, M. R. Hicks, A. Rodger, D. E. Lopez-Perez, G. Revilla-Lopez and C. Aleman, *Soft Matter*, 2011, **7**, 11405–11415.
61. J. D. Hartgerink, E. Beniash and S. I. Stupp, *Science*, 2001, **294**, 1684–1688.
62. J. D. Hartgerink, E. Beniash and S. I. Stupp, *Proc. Natl. Acad. Sci. U. S. A.*, 2002, **99**, 5133–5138.
63. D. A. Harrington, E. Y. Cheng, M. O. Guler, L. K. Lee, J. L. Donovan, R. C. Claussen and S. I. Stupp, *J. Biomed. Mater. Res., Part A*, 2006, **78**, 157–167.
64. R. N. Shah, N. A. Shah, M. M. Del Rosario Lim, C. Hsieh, G. Nuber and S. I. Stupp, *Proc. Natl. Acad. Sci. U. S. A.*, 2010, **107**, 3293–3298.
65. A. Mata, L. Hsu, R. Capito, C. Aparicio, K. Henrikson and S. I. Stupp, *Soft Matter*, 2009, **5**, 1228–1236.
66. A. Mahler, M. Reches, M. Rechter, S. Cohen and E. Gazit, *Adv. Mater.*, 2006, **18**, 1365–1370.
67. V. Jayawarna, M. Ali, T. A. Jowitt, A. F. Miller, A. Saiani, J. E. Gough and R. V. Ulijn, *Adv. Mater.*, 2006, **18**, 611–614.
68. V. Jayawarna, S. M. Richardson, A. R. Hirst, N. W. Hodson, A. Saiani, J. E. Gough and R. V. Ulijn, *Acta Biomater.*, 2009, **5**, 934–943.
69. M. Zhou, A. M. Smith, A. K. Das, N. W. Hodson, R. F. Collins, R. V. Ulijn and J. E. Gough, *Biomaterials*, 2009, **30**, 2523–2530.
70. T. Liebmann, S. Rydholm, V. Akpe and H. Brismar, *BMC Biotechnol.*, 2007, **7**, 88.
71. S. G. Elner and V. M. Elner, *Invest. Ophthalmol. Visual Sci.*, 1996, **37**, 696–701.
72. E. Ruoslahti and M. D. Pierschbacher, *Science*, 1987, **238**, 491–497.
73. R. Ortega-Velázquez, M. L. Díez-Marqués, M. P. Ruiz-Torres, M. González-Rubio, M. Rodríguez-Puyol and D. R. Puyol, *FASEB J.*, 2003, **17**, 1529–1531.
74. S. P. Massia and J. A. Hubbell, *J. Cell Biol.*, 1991, **114**, 1089–1100.

75. Y. Jin, X.-D. Xu, C.-S. Chen, S.-X. Cheng, X.-Z. Zhang and R.-X. Zhuo, *Macromol. Rapid Commun.,* 2008, **29**, 1726–1731.
76. M. O. Guler, L. Hsu, S. Soukasene, D. A. Harrington, J. F. Hulvat and S. I. Stupp, *Biomacromolecules,* 2006, **7**, 1855–1863.
77. B. Lauweryns, J. J. van den Oord, R. Volpes, B. Foets and L. Missotten, *Invest. Ophthalmol. Visual Sci.,* 1991, **32**, 2079–2085.
78. A. V. Ljubimov, R. E. Burgeson, R. J. Butkowski, A. F. Michael, T. T. Sun and M. C. Kenney, *Lab. Invest.,* 1995, **72**, 461–473.
79. R. M. Gouveia, V. Castelletto, S. G. Alcock, I. W. Hamley and C. J. Connon, *J. Mater. Chem. B,* 2013, **1**, 6157–6169.
80. M. J. Webber, J. Tongers, M.-A. Renault, J. G. Roncalli, D. W. Losordo and S. I. Stupp, *Acta Biomater.,* 2010, **6**, 3–11.
81. R. R. Jones, V. Castelletto, C. J. Connon and I. W. Hamley, *Mol. Pharmaceutics,* 2013, **10**, 1063–1069.
82. W. Helen, C. L. Merry, J. J. Blaker and J. E. Gough, *Biomaterials,* 2007, **28**, 2010–2020.
83. G. Cheng, V. Castelletto, R. R. Jones, C. J. Connon and I. W. Hamley, *Soft Matter,* 2011, **7**, 1326–1333.
84. L. H. Lee, R. Peerani, M. Ungrin, C. Joshi, E. Kumacheva and P. Zandstra, *Stem Cell Res.,* 2009, **2**, 155–162.
85. X. Liu, J. Y. Lim, H. J. Donahue, R. Dhurjati, A. M. Mastro and E. A. Vogler, *Biomaterials,* 2007, **28**, 4535–4550.
86. E. Y. L. Waese and W. L. Stanford, *Stem Cell Res.,* 2011, **6**, 34–49.
87. J. Z. Gasiorowski, S. J. Liliensiek, P. Russell, D. A. Stephan, P. F. Nealey and C. J. Murphy, *Biomaterials,* 2010, **31**, 8882–8888.
88. A. J. Engler, S. Sen, H. L. Sweeney and D. E. Discher, *Cell,* 2006, **126**, 677–689.
89. A. Engler, L. Bacakova, C. Newman, A. Hategan, M. Griffin and D. Discher, *Biophys. J.,* 2004, **86**, 617–628.
90. C.-M. Lo, H.-B. Wang, M. Dembo and Y.-L. Wang, *Biophys. J.,* 2000, **79**, 144–152.
91. M. G. Haugh, C. M. Murphy, R. C. McKiernan, C. Altenbuchner and F. J. O'Brien, *Tissue Eng., Part A,* 2010, **17**, 1201–1208.
92. R. J. Pelham and Y.-L. Wang, *Proc. Natl. Acad. Sci. U. S. A.,* 1997, **94**, 13661–13665.
93. E. Hadjipanayi, V. Mudera and R. A. Brown, *J. Tissue Eng. Regener. Med.,* 2009, **3**, 77–84.
94. B. Harland, S. Walcott and S. X. Sun, *Phys. Biol.,* 2011, **8**, 1478–3975.
95. D. E. Jaalouk and J. Lammerding, *Nat. Rev. Mol. Cell Biol.,* 2009, **10**, 63–73.
96. A. Petersen, P. Joly, C. Bergmann, G. Korus and G. N. Duda, *Tissue Eng., Part A,* 2012, **18**, 1804–1817.
97. R. A. Marklein and J. A. Burdick, *Adv. Mater.,* 2010, **22**, 175–189.
98. M. P. Lutolf and J. A. Hubbell, *Nat. Biotechnol.,* 2005, **23**, 47–55.
99. M. P. Lutolf and H. M. Blau, *Adv. Mater.,* 2009, **21**, 3255–3268.

100. A. J. Keung, S. Kumar and D. V. Schaffer, *Annu. Rev. Cell Dev. Biol.*, 2010, **26**, 533–556.
101. J. H. Collier, J. S. Rudra, J. Z. Gasiorowski and J. P. Jung, *Chem. Soc. Rev.*, 2010, **39**, 3413–3424.
102. E. S. Place, N. D. Evans and M. M. Stevens, *Nat. Mater.*, 2009, **8**, 457–470.
103. M. Arnold, E. A. Cavalcanti-Adam, R. Glass, J. Blummel, W. Eck, M. Kantlehner, H. Kessler and J. P. Spatz, *ChemPhysChem*, 2004, **5**, 383–388.
104. C. Selhuber-Unkel, T. Erdmann, M. Lopez-Garcia, H. Kessler, U. S. Schwarz and J. P. Spatz, *Biophys. J.*, 2010, **98**, 543–551.
105. J. A. Deeg, I. Louban, D. Aydin, C. Selhuber-Unkel, H. Kessler and J. P. Spatz, *Nano Lett.*, 2011, **11**, 1469–1476.
106. G. Maheshwari, G. Brown, D. A. Lauffenburger, A. Wells and L. G. Griffith, *J. Cell Sci.*, 2000, **113**, 1677–1686.
107. M. J. Wilson, S. J. Liliensiek, C. J. Murphy, W. L. Murphy and P. F. Nealey, *Soft Matter*, 2012, **8**, 390–398.
108. F. Yang, C. G. Williams, D.-A. Wang, H. Lee, P. N. Manson and J. Elisseeff, *Biomaterials*, 2005, **26**, 5991–5998.
109. S. Q. Liu, R. Tay, M. Khan, P. L. R. Ee, J. L. Hedrick and Y. Y. Yang, *Soft Matter*, 2010, **6**, 67–81.
110. C. N. Salinas and K. S. Anseth, *Biomaterials*, 2008, **29**, 2370–2377.
111. U. Hersel, C. Dahmen and H. Kessler, *Biomaterials*, 2003, **24**, 4385–4415.
112. D. A. Bettinger, D. R. Yager, R. F. Diegelmann and I. K. Cohen, *Plast. Reconstr. Surg.*, 1996, **98**, 827–833.
113. E. K. Purcell, Y. Naim, A. Yang, M. K. Leach, J. M. Velkey, R. K. Duncan and J. M. Corey, *Biomacromolecules*, 2012, **13**, 3427–3438.
114. B. M. Gillette, J. A. Jensen, M. Wang, J. Tchao and S. K. Sia, *Adv. Mater.*, 2010, **22**, 686–691.
115. J. S. Mohammed and W. L. Murphy, *Adv. Mater.*, 2009, **21**, 2361–2374.
116. D. M. Le, K. Kulangara, A. F. Adler, K. W. Leong and V. S. Ashby, *Adv. Mater.*, 2011, **23**, 3278–3283.
117. M. A. Schwartz, *Cold Spring Harbor Perspect. Biol.*, 2010, **2**, a005066.
118. V. Vogel and M. Sheetz, *Nat. Rev. Mol. Cell Biol.*, 2006, **7**, 265–275.
119. C. Lui, K. Lee and C. M. Nelson, *Biomech. Model. Mechanobiol.*, 2012, **11**, 1241–1249.
120. A. M. Shewan, M. Maddugoda, A. Kraemer, S. J. Stehbens, S. Verma, E. M. Kovacs and A. S. Yap, *Mol. Biol. Cell*, 2005, **16**, 4531–4542.
121. S. Dupont, L. Morsut, M. Aragona, E. Enzo, S. Giulitti, M. Cordenonsi, F. Zanconato, J. Le Digabel, M. Forcato, S. Bicciato, N. Elvassore and S. Piccolo, *Nature*, 2011, **474**, 179–183.
122. T. Wakatsuki, R. B. Wysolmerski and E. L. Elson, *J. Cell Sci.*, 2003, **116**, 1617–1625.
123. Y. Cai, N. Biais, G. Giannone, M. Tanase, G. Jiang, J. M. Hofman, C. H. Wiggins, P. Silberzan, A. Buguin, B. Ladoux and M. P. Sheetz, *Biophys. J.*, 2006, **91**, 3907–3920.

124. M. Vicente-Manzanares, X. Ma, R. S. Adelstein and A. R. Horwitz, *Nat. Rev. Mol. Cell Biol.*, 2009, **10**, 778.
125. D. E. Discher, P. Janmey and Y. L. Wang, *Science*, 2005, **310**, 1139–1143.
126. M. Chrzanowska-Wodnicka and K. Burridge, *J. Cell Biol.*, 1996, **133**, 1403–1415.
127. B. Zhao, L. Li and K. L. Guan, *J. Cell Sci.*, 2010, **123**, 4001–4006.
128. J. H. Hong, E. S. Hwang, M. T. McManus, A. Amsterdam, Y. Tian, R. Kalmukova, E. Mueller, T. Benjamin, B. M. Spiegelman, P. A. Sharp, N. Hopkins and M. B. Yaffe, *Science*, 2005, **309**, 1074–1078.
129. D. Berg, C. Holzmann and O. Riess, *Nat. Rev. Neurosci.*, 2003, **4**, 752–762.
130. S. Basu, N. F. Totty, M. S. Irwin, M. Sudol and J. Downward, *Mol. Cell*, 2003, **11**, 11–23.
131. F. Ren, L. Zhang and J. Jiang, *Dev. Biol.*, 2010, **337**, 303–312.

CHAPTER 5

Microencapsulation of Probiotic Bacteria into Alginate Hydrogels

M. T. COOK[*,a], D. CHARALAMPOPOULOS[b], AND
V. V. KHUTORYANSKIY[a]

[a] School of Pharmacy, University of Reading, Reading RG6 6AD, UK;
[b] Department of Food and Nutritional Sciences, University of Reading, Reading RG6 6AD, UK
*E-mail: M.T.Cook@Reading.ac.uk

5.1 Introduction

Probiotic bacteria have become incredibly successful in recent years, with some reports predicting a global market value of up to 32.6 billion US dollars. Dairy beverages account for the largest share, but there is a predicted rise in the number of niche applications, such as in chocolate or ice cream. However, currently the ability of these microorganisms to survive in the gastrointestinal (GI) tract after oral administration or during storage in any foods is highly variable, with a dependency on the strain used. As a result, strains may often be selected for use based on their ability to survive in these conditions rather than their ability to exert a positive effect on the host. However, the crossover of pharmaceutics with functional food research may be able to alter this. The entrapment of probiotic cells into polymer matrices, termed 'microencapsulation', has become an effective method of improving the survival of these cells both during storage and during GI transit. This process is a challenging one for a research scientist, principally because of the sensitivity of the cells and the need to make a product with a composition suitable for repeated human ingestion in large quantities. The most popular method of microencapsulating probiotics is to entrap them into ionic hydrogels of a natural polysaccharide, alginate. These systems have been shown to have varying efficacy in protecting probiotics, as evaluated by a number of groups. This chapter primarily concerns the currently available research in the field of alginate encapsulation applied to probiotic cells. Modifications to this system most commonly occur as coats or by incorporation of another polymer into the alginate matrix, which will also be discussed. The key publications in this field will be discussed where necessary.

How probiotics are believed to act is perhaps best explained through the history of these microorganisms. Ilya Ilyich Metchnikoff, born in 1845 in what is now Ukraine, could be credited with starting the sequence of events leading to the discovery of probiotics.[1] Metchnikoff was an interesting man who had a troubled life, in which he saw the tragic death of three brothers (the eldest of whom, Ivan Ilyich, was to become the subject of a now-famous novella by Leo Tolstoy). The death of his first wife and mother led, in part, to two failed suicide attempts. He eventually became financially independent through inheritance and went on to make the most important findings of his career. His study of starfish larvae led to the discovery of phagocytosis, and later its role in immunity. For this he would win a Nobel prize in 1908. The work of importance to us, however, occurred later in his life. As he aged and became unwell, Metchnikoff started a new field of study, gerontology, which he devoted himself to. This study led him to the belief that ageing was related to putrefaction of the intestine, and, in his book *The Prolongation of Life* (originally published in Russian in 1906), outlined his theory that lactic acid bacillus are responsible for the ageing process. Metchnikoff's belief that probiotics could extend life was adopted by Minoru Shirota of Kyoto University, Japan. His research led him to the belief that there were bacteria with particularly high levels of lactic acid production in the vagina and that it 'has been found that if such bacteria enter the digestive system of an infant they grow therein... and thus protect the intestines of the infant from pathogenic microbes.'[2] It is clear that his proposed mechanism for action of these cells is by competition with pathogens, resulting in a reduced risk of infection. Shirota's specific strain, *Lactobacillus casei* Shirota, was isolated after incubation of lactic acid bacteria obtain from infants faeces in acid at pH 3.1–3.3 for up to 24 h. This eliminated all bacteria other than *Lactobacillus casei*. However, not all probiotics show this degree of tolerance to acid and may be killed by the low pH of the stomach. Since Shirota's research there have been great advances in the number of probiotics and research into their likely mechanisms of action. There are now numerous bacteria of the genera *Lactobacillus* and *Bifidobacterium* currently sold as probiotics as well as a yeast, *Saccharomyces boulardii*. Their mechanisms of action are still fairly poorly understood, but are believed be a result of two processes: the presentation of surface molecules to the host's immune system or the production of beneficial compounds *in vivo*.[3] These surface molecules or secretions have been claimed to alleviate numerous diseases, but the strongest clinical data lies in the treatment of traveller's diarrhoea, pouchitis and inflammatory bowel disease. This mixture of diseases and the uncertainty in the exact mechanism of action makes the delivery of these cells a difficult task. Generally speaking, however, the site of action is the large intestine, where these 'commensal' microbes usually reside, and the clear pharmaceutical challenge is the protection of the more sensitive of these species in the stomach, and delivery of encapsulated cells to the large intestine.

5.2 The Chemistry of Alginates

The structure and functionality of alginates has been extensively studied by many groups, but the bulk of the research can be attributed to a handful of Norwegian researchers who have been prolific in their characterization of alginates and ionic alginate hydrogels. The most prolific of this group is Olav Smidsrod, who has published over 70 papers on alginate, most commonly in collaboration with Gudmund Skjåk-Bræk. Fantastically detailed discoveries and methods for the characterization of alginate biomaterials may be sourced through their work. This work includes the characterization of alginate through hydrogen[4] and carbon[5] NMR, study of the relationship between the alginate structure and gel properties,[6] and the study of the coating of alginate hydrogels with cationic polymers.[7] Much of what is known about alginate-based materials can be attributed to members of this group, making them an invaluable source of information on the material.

Alginates are polysaccharides produced naturally in the cell walls of algae, meaning that these materials can be sustainably produced. As the polysaccharide is a naturally occurring material, the exact structure of alginates varies dependent upon the precise source of the polymer. The monomers along the alginate macromolecule have carboxylic acid functionality and, as a result, any charged carboxylic groups can associate to cations in solution. This association can lead to crosslinking of the macromolecules if certain divalent metal cations are used, leading to 'ionic' gelation of the polymer. Chemically speaking, alginate is a copolymer of β-D-mannuronic acid (M) and α-L-guluronic acid (G), joined by 1 → 4 glycosidic linkages. These M and G residues are 'randomly' arranged as GG, MM, GM or MG blocks (Figure 5.1), though the order is consistent for alginates from the same source. The ratio of M and G blocks is the most important variable between different alginates and greatly affects the polymer's properties. This M/G ratio can be determined experimentally by the depolymerization of the alginate, followed by the selective removal of M and G blocks from solution. This selective removal of monomers is achieved by lowering the pH of solution, causing the acids to 'crash out' of solution once their pK_a is passed. M and G have slightly different pK_a values (3.38 and 3.65,[8] respectively), making this possible. Although M and G are simply c5 epimers, they

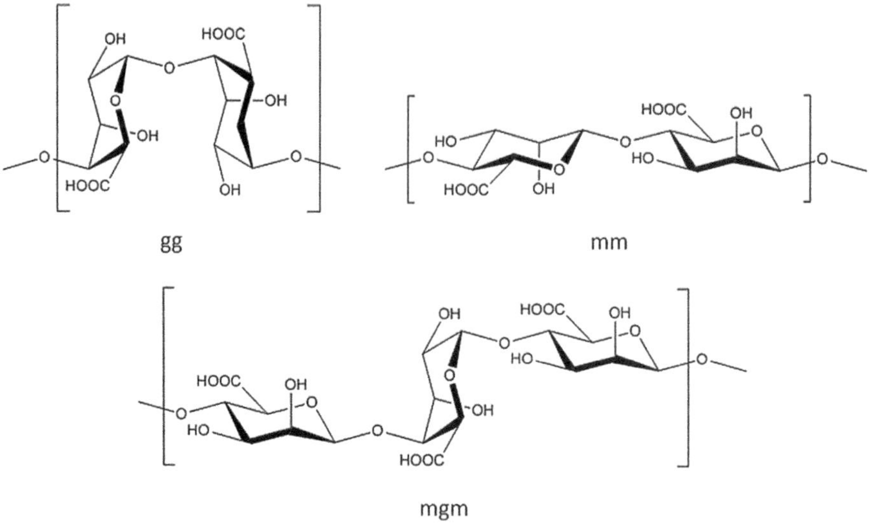

Figure 5.1 The key repeat units of alginates.

have markedly different properties. M and G residues in alginates can both associate to divalent metal cations but with varying binding strengths. The most commonly used metal crosslinker for alginates is calcium, which forms crosslinks between blocks of G residues in the alginate. The complex formed between divalent calcium and G residues is named, due to its appearance, an 'egg-box' structure. This structure allows multiple sites for interaction between the electron-rich carboxylic groups and the electron-deficient calcium, making the complex very strong. The differing ability of M and G to bind to divalent metals has been examined in great depth, with different binding motifs associated with different metals. For instance, it has been shown that Ca^{2+} binds to GG and MG blocks, Ba^{2+} to G and M blocks, and Sr^{2+} to G blocks only.[9] Once crosslinked under the right conditions, alginates can form gels of varying strength. These materials are well suited to the encapsulation of sensitive cells because there is no need for a covalent crosslinker which may harm the cells during the production of these materials. The crosslinks in the gel may be removed either by competition for binding with monovalent cations or by sequestration of the metal by chelators which compete with alginate for metal complexation. For Ca^{2+} these chelators are typically phosphates or citrates. The removal of crosslinks from the gel results in the dissolution of the gel matrix, and release of encapsulated cells.

Alginate contains carboxylate residues on its G and M monomers. A result of this functionality is that as pH is reduced to below the pK_a of the monomers (3.38 and 3.65 for M and G, respectively) the carboxylate groups become protonated to a greater degree, neutralizing their charge. This protonation leads to the aggregation of alginate polymers, and the formation of an 'acid–gel'. This acid–gel character has been extensively studied by Draget et al.,[10] and appears to start at around pH 4 and increase in hardness up to pH 2. This means that alginate hydrogels are resistant to dissolution in the stomach, even in the presence of Ca^{2+} chelators. After passage through the stomach *in vivo*, the cells then pass into the duodenum of the small intestine, at which point the pH rises to approximately pH 6.0, and the acid–gel character of the material is lost. The hydrogel may then dissolve by the aforementioned sequestration or monovalent salt competition. A reflection of the acid–gel character described can be seen when alginate is purchased. Alginates are available as either the acid form or the sodium salt, with the former having a much lower solubility.

Another important consideration of alginate hydrogels is their permeability. If the hydrogel produced is too porous, it is likely that there will be loss of encapsulated load by diffusion through these pores. Additionally, if the materials are permeable to proteins it is possible that some enzymatic damage may occur to the cell. Precise measurement of gel porosity is a relatively difficult task and will be dependent on the properties of the specific alginate grade and the production conditions of the hydrogels. However, a general size range for these pores has been given by Andresen et al.[11] as 5–200 nm in diameter. These pores are too small to allow passage of any encapsulated probiotic cells, which should result in retention of the cells within the hydrogel matrix until the breakdown of the network. This size range does, however, allow the passage of some proteins. This passage is likely to be pH dependent, as well as size dependent, due to the aggregation of alginate macromolecules at low pH. Thus, when studying these materials, impermeability of the matrix must not be assumed. The passage of molecules into the hydrogel matrix is usually easily determined by standard quantification techniques. An example of this is demonstrated by Mørch et al.,[9] who showed that the permeability of these materials to immunoglobulin G (a small, 150 kDa, antibody) can vary according to both the cation used to gel the alginate and the M/G ratio of the alginate used. Most importantly, the high-M content alginate appeared to be less permeable to immunoglobulin G than the corresponding high-G alginate. This was a result of the higher effective concentration of alginate in high-M alginate gels, due to their reduced size relative to high-G alginate gels. However, it should be noted that this effect was dependent on the cation used.

5.3 Producing Alginate Hydrogels

There are various methods that can be used to produce alginate gels, which can be broadly separated into those that use extrusion or emulsion, and those that use internal or external gelation. The former two terms describe the processes by which droplets of alginate are created, the latter two describe the methods by which these droplets may be crosslinked to form a hydrogel. Extrusion (Figure 5.2) describes the formation of droplets by the dripping of an alginate–cell solution through a 'cavity' into crosslinking

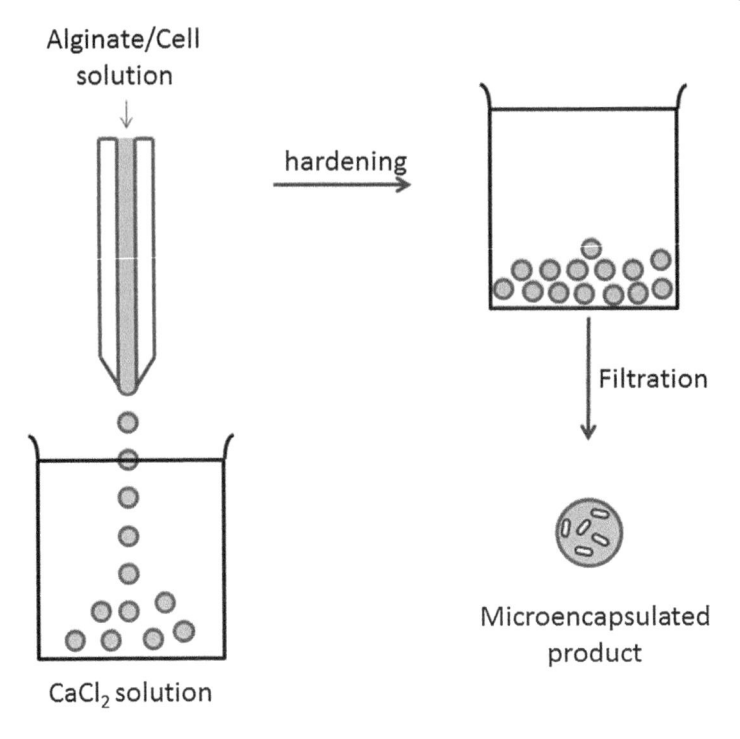

Figure 5.2 Overview of the extrusion method for producing alginate microcapsules.

solution. This cavity is commonly simply a needle, for benchtop experiments, using a syringe as a source of positive pressure. The size of the droplets created, and therefore the approximate size of the final microcapsules, is dependent on the diameter of the hole through which the alginate is extruded. This results in the formation of large matrices of approximately 1 mm size when using a 21 gauge needle. The size of these hydrogels can be reduced by using syringes of higher gauges, but to produce very small microcapsules requires the use of more specialized equipment. There are a number of tools that can be used to produce these microcapsules, most of which use specialized nozzles to allow the production of much finer spray. Solutions of alginate may then be sprayed directly into crosslinking solution in order to produce alginate microcapsules.

The emulsion method of droplet formation proceeds through the formation of a water-in-oil emulsion, in which the discrete water phase contains an alginate–cell mixture. The crosslinking of alginate is then achieved by either the addition of a large volume of calcium chloride solution (to produce Ca–alginate matrices) or by internal gelation, as discussed later. The initial emulsion is usually stabilized by a surfactant, the choice of which alters the final morphology of the capsules (Figure 5.3).[12]

Both methods of droplet formation have advantages and disadvantages, and choosing one over the other is down to the researcher's discretion.

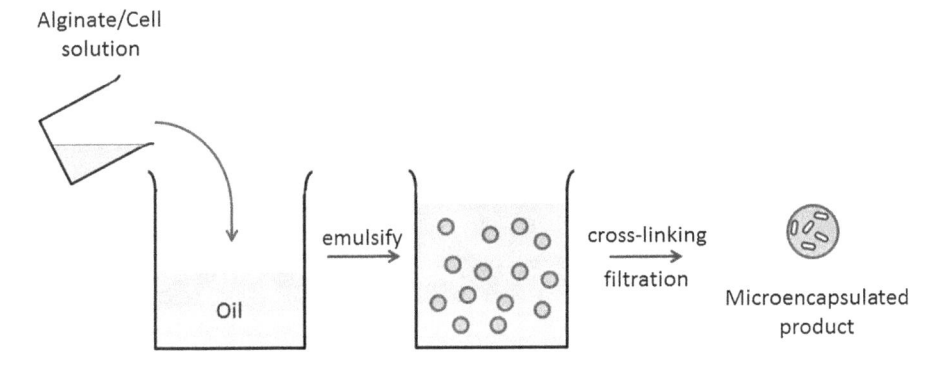

Figure 5.3 The emulsion method of alginate microcapsule production.

In simple laboratory experimentation, emulsion allows for the formation of very small microcapsules (around 100 µm) without the need for special tools, as in extrusion. The extrusion technique, however, allows the formation of a very homogeneous product with easy tailoring of microcapsule size based on the diameter of the cavity through which the alginate is passed. Conversely, the products from the emulsion process results in a very variable product, with many factors determining the final microcapsule size, such as emulsion stirring speed, surfactant choice, surfactant concentration, and oil/water phase volumes. Additionally, the emulsion procedure has the potential to leave oil in the final product which could prove difficult to remove.

The gelation of alginate may be achieved by one of two methods, termed 'external' and 'internal' gelation. External gelation refers to the crosslinking of alginate by divalent metals by the diffusion of the metals into an alginate droplet. This gelation mechanism is most common for the production of alginate microcapsules by the extrusion method as the alginate is extruded into a divalent metal solution which diffuses into the droplet. As gelation proceeds at the extremities of the droplet, in the first instance there is a reduction in the volume of the crosslinked alginate macromolecules which in turn reduces the effective concentration in the remainder of the droplet. This process continues during gelation and, as a result, a heterogeneous hydrogel is produced, with a higher concentration of alginate at the periphery. This has been elegantly visualized by Mørch et al.[9] by the labelling of alginate and subsequent visualization of the microbeads produced by confocal microscopy. It was demonstrated that the homogeneity of the hydrogel formed *via* extrusion and external gelation is dependent on the gelling solution used (Figure 5.4). It appears that an increase in the ionic strength of the gelling medium allows for a more homogeneous distribution of alginate throughout the microcapsule. This is believed to be a result of the greater concentration gradient between the inside of the alginate droplet and the metal solution driving the ions into the alginate to a greater extent. Additionally, the species of cation used appears to alter the homogeneity of the material, which was attributed to the different strengths of binding

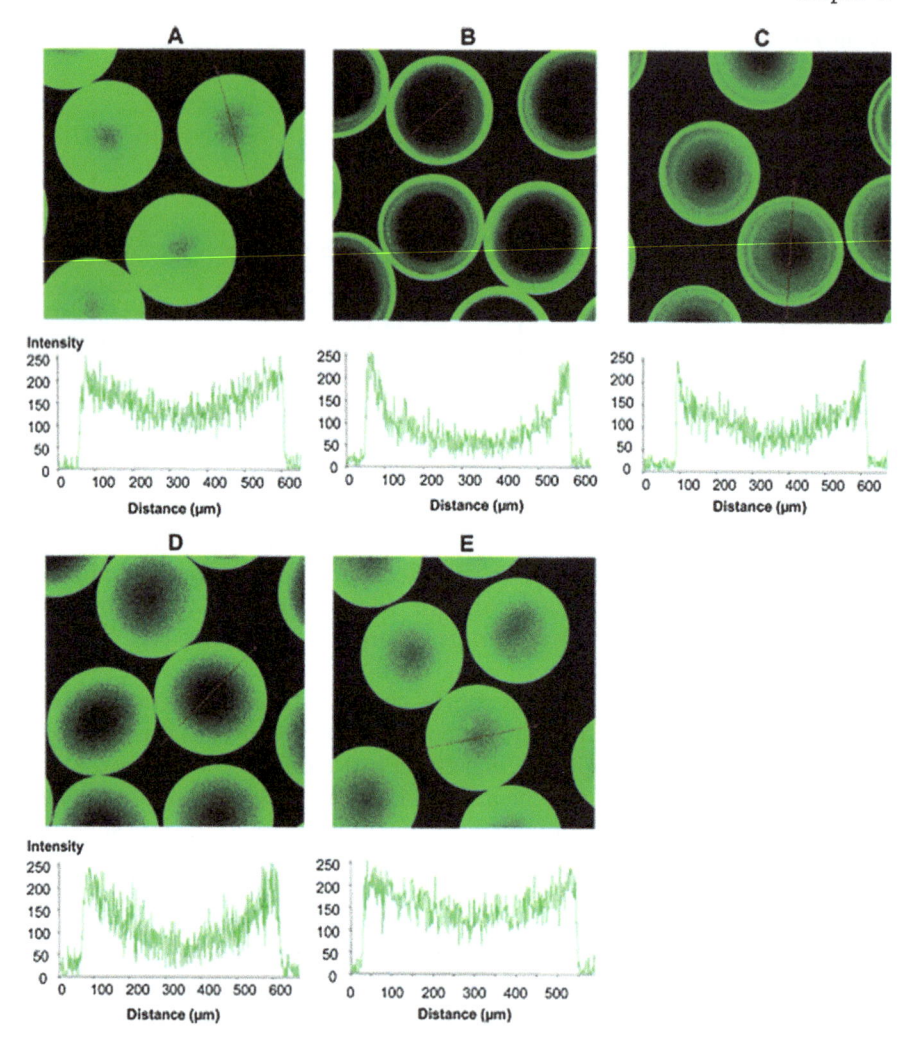

Figure 5.4 Heterogeneous alginate gels visualized by confocal microscopy. Gelling conditions were (all in 0.15 M mannitol): (A) 50 mM CaCl$_2$, (B) 10 mM BaCl$_2$, (C) 20 mM BaCl$_2$, (D) 20 mM CaCl$_2$, (E) 50 mM SrCl$_2$. Graphs below images show the pixel intensity across the microcapsule diameter. From ref. 9, with permission. Copyright American Chemical Society 2006.

between the polymer and the various cations. The external gelation method may also be used in conjunction with the emulsion process, by addition of a large volume of crosslinker after the production of the initial emulsion.

Internal gelation refers to a very different mechanism of gel formation. It involves the suspension of an 'inert' form of crosslinker (with respect to gelation) in the alginate–cell droplet and the addition of another compound which reacts with the inert crosslinker to activate it. This is typically achieved by the suspension of insoluble calcium carbonate into the alginate droplet, followed by the addition of acetic acid. The acetic acid reacts with the calcium carbonate to form carbonic acid, which is removed from solution as carbon dioxide gas. This results in the formation of calcium crosslinks at the site of reaction, and the production of an alginate hydrogel. This method of gelation is most commonly used in conjunction with the emulsion process, as acetic acid is able to pass through the oil phase, into the alginate droplets with ease. The gels produced by this method are more homogeneous than those produced by the external gelation mechanism,[13] but the addition of acetic acid to the mixture may be detrimental to some strains of probiotic. An alternative method of liberating active crosslinker to internally gel alginate has been described by Draget *et al.*[14] In this method, calcium carbonate was suspended in alginate solution as before, but with the incorporation of glucono-δ-lactone. Glucono-δ-lactone is a simple sugar which slowly hydrolyses in solution to form gluconic acid. This hydrolysis results in a gradual reduction in solution pH until a point at which carbonic acid may be formed from the suspended calcium carbonate, releasing active calcium ions to induce alginate gelation. The rate of glucono-δ-lactone hydrolysis, and therefore onset of gelation, may be altered by changing the temperature of the solution.

5.4 Protecting Probiotics—Demonstrating Efficacy of Alginate Microcapsules

We now turn to the application of these materials. As touched upon in the Introduction, the majority of research focuses on the protection of cells from acid. These acidic environments are either gastric, *i.e.* the stomach, or food-based, *e.g.* yogurt. In either of these environments the pH is sufficiently low to cause damage to probiotic cells, reducing the efficacy of any supplement. The precise conditions in each, however, are very different. For gastric pH challenge, cells will typically be exposed to a very acidic (pH 1.5–2.0) solution for a few hours at 37 °C. In foodstuffs, however, cells will be exposed to a less harsh pH, for much longer durations. For instance, yogurt only has a pH of around 4.6,[15] but will need to be stored for 2–3 weeks under refrigeration (4 °C). These two environments pose considerably different challenges, but alginate encapsulation has demonstrated success in improving the survival of microorganisms in each. Normally during these processes the number of live (viable) cells will be quantified using either serial dilution and plating techniques or live/dead cell staining. For those unfamiliar with the techniques, plate counting is often seen as the least ambiguous method of cell enumeration. Plate counting, often called viable counting, initially involves the preparation of a dilution series from the sample to be tested in a suitable medium such as phosphate buffered saline, in which the ionic strength of the solution is 'physiological' (154 mM) so as not to harm the cells. This dilution series is then spread onto the surface of a solid agar growth medium, which can either be non-selective, allowing the growth of all cells, or selective, only allowing the growth of certain bacteria. These agar plates will be incubated either aerobically or anaerobically, depending on the cells, until bacteria colonies appear on the plates. These colonies can have various morphologies, depending on the strain, and are the product of a single cell's reproduction by binary fission. Thus, by counting the number of colonies present on the plate, and multiplying by the dilution factor one can determine the number of microbes present in the sample which are viable and capable of reproduction. An alternative method is live/dead staining of the sample. This involves the exposure of a sample of cells to two dyes, often SYTO-9 and propidium iodide. The ability of these dyes to stain cells depends on the integrity of the cell membrane. SYTO-9 (or another general stain) freely permeates the cell membrane, dying every cell, but the propidium iodide can only pass into cells with a damaged membrane. This allows differentiation between 'live' cells (those with intact membranes) and 'dead' cells (those without). The number of these 'live' or 'dead' cells may be determined by techniques such as fluorescence microscopy or flow cytometry. It is important to note that these two techniques measure different things, and that cells that are 'live' by staining may not be 'live' by plate counting; a result of the membrane being intact but the cell being unable to reproduce due to some other damage.

For gastrointestinal delivery, much work has been conducted *in vitro*, and there are some *in vivo* studies, in animals and humans. Examples of *in vitro* study include the work of Chandramouli *et al.*[16] in which two strains of *Lactobacillus acidophilus* were examined for their ability to survive before and after encapsulation during incubation in a complex milk-based medium intended to simulate the conditions in the stomach. Cells were first encapsulated by an extrusion technique with external gelation, using a specialized spray device. Exposure to the simulated gastric solution led to a decrease in numbers of viable cells, but the viability of the probiotics was higher after alginate encapsulation. During 3 h of incubation in this pH 2.0 media, encapsulation gave a 100-fold increase in numbers of viable cells for both strains used. This study also yielded some important additional findings on the effect of some processing parameters on the efficacy of alginate microcapsules. It was found that increasing the size of the microcapsules and the concentration of alginate extruded improved the survival of cells in acid at pH 2. This effect of increasing alginate concentration improving cell survival in acid was also seen by Lee and Heo.[17] The mechanism by which these materials protect cells is often hypothesized to be a result of buffering in the local environment around the cells, lowering the activity of the acid. This hypothesis is consistent with Lee and Heo's finding, in so far as increasing alginate concentration will increase the buffering capacity of the solution and therefore lower activity of acid to a greater extent. It has also been found by Cook *et al.*[18] that alginate hydrogels may be dehydrated and still retain their ability to protect probiotic *Bifidobacterium breve* from low pH. Once the stomach has been passed, it is likely that the alginate hydrogels can dissolve, releasing their load. Due to the size of the encapsulated cells, diffusion as a mechanism of release is unlikely.

The pH within alginate gels during exposure to acid has been studied by Cook et al.[19] This was conducted by labelling a probiotic *Bifidobacterium* strain, *B. breve*, with two pH-sensitive fluorophores. Confocal microscopy of alginate microcapsules containing these labelled cells during acid exposure allowed the visualization of the pH within the materials by a ratiometric approach (Figure 5.5). This revealed that the pH within the microcapsules decreased from the periphery first, and that coating with chitosan slowed the penetration of acid. This lowering of pH should also coincide with the formation of an acid–gel, lowering permeability. This data implies that the protective effect of these systems is a result of the decreased exposure time of the cells to low pH. Though this does not allow complete elucidation of the method, it does not contradict the hypothesis that buffering in the local environment raises the pH above that of the surrounding medium.

In vivo trials of these alginate encapsulated probiotics are relatively limited. Graff et al.[20] found that feeding Wistar rats encapsulated probiotic yeast had increased numbers of viable probiotics found in their faeces than those fed the yeast alone, though it should be noted that the enteric polymer Eudragit was incorporated into the alginate gels. A trial by Cui et al.[21] of modified alginate microcapsules in human subjects saw that encapsulation led to an 11.5–30 fold improvement in the survival of a probiotic *Bifidobacterium* during gastric passage. However, the starting cell numbers for the control and encapsulated product were not identical, making the study not as rigorous as would be required to draw very strong conclusions from this. Although these *in vivo* studies are limited they do provide some hope that these alginate encapsulation systems have potential to improve the efficacy of probiotic supplements.

The efficacy of any microencapsulation system in protecting probiotic bacteria in foodstuffs is far easier to determine than for GI protection. As only the evaluation of survival within the foodstuff is required, the researcher may

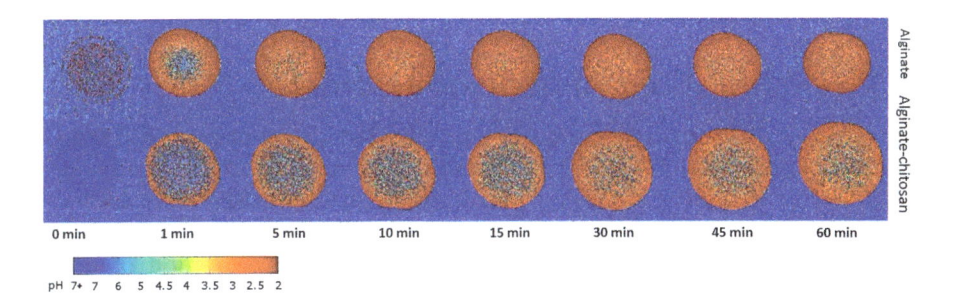

Figure 5.5 pH environments within alginate and chitosan-coated alginate microcapsules during exposure to simulated gastric solution (pH 2.0) and incubation (60 min, 37 °C). From ref. 19, with permission. Copyright American Chemical Society 2013.

collect all data needed without the need for *in vivo* evaluation. For example, Kailasapathy[22] has demonstrated the survival of *Lactobacillus* and *Bifidobacterium* strains in yogurt during 7 weeks of storage. This study included the testing of both free cells and those encapsulated into alginate microcapsules produced by the emulsion method. These alginate microcapsules had a starch 'filler' incorporated, to improve the structure of the microcapsules. It was found that the microencapsulation within this system improved the survival of *Lactobacillus acidophilus* 100-fold and *Bifidobacterium lactis* 10-fold over 7 weeks of storage with refrigeration. During this storage period the pH of the yogurt decreased to pH 4.25. It was seen that during this period, encapsulated cells were still 'active' in the hydrogel, and associated cell processes could be monitored, *e.g.* polysaccharide production and acidification. The study also included a survey of the sensory properties of these yogurts. This sensory evaluation included yogurt appearance and colour, body and texture, acidity, flavour, after taste and the overall enjoyment of the yogurt. Results from this study were very positive overall, with no significance found in the majority of categories. However, it was found that there was a significant difference in the texture of the yogurts containing microencapsulated bacteria. The microcapsules used in this study were around 300 µm in size and a reduction in microcapsule volume might improve the texture of the yogurt, although it might be associated with a decreased ability to protect the probiotics.

Although much research has been conducted on the microencapsulation of probiotics into alginate hydrogels, it should be noted that not all studies show this system to be effective. For example, Truelstrup Hansen *et al.*[23] found that the microencapsulation of several probiotic *Bifidobacterium* strains into alginate microcapsules produced by emulsion with external gelation was not able to offer significantly enhanced viability during exposure to simulated gastric conditions. A possible reason for this is the very small size (20–70 µm) of microcapsule used. It has previously been suggested that size plays an important role in the ability of these microcapsules to protect probiotic cells, so it may be that these microcapsules were below a desirable size limit.

5.5 Modifications of Alginate Hydrogels

Despite their demonstrated efficacy in some areas, alginate hydrogels as microencapsulation matrices for probiotic bacteria have some shortcomings. The greatest of these are the loss of protection with size reduction and non-specificity of cell delivery in the intestine. Additionally, while most studies found alginate microcapsules to be an effective method of protecting cells from acid, the effect could still be improved to deliver more viable cells. Modification of alginate hydrogels is usually not attempted by chemical modification of the alginate itself, because of issues surrounding potential toxicity and the need for new certification by the appropriate food or pharmaceutical bodies. Typically, microcapsules have been modified by coating the matrix with other materials or by mixing additional excipients into the gel matrix. Krasaekoopt et al.[24] have evaluated the potential of some coating materials to improve the survival of probiotic microorganisms in simulated GI conditions. Coating with chitosan, poly-L-lysine–alginate and alginate was attempted by simply dipping the microcapsules into solutions of the polymers, and many groups have used this approach. At certain pHs alginate is anionic whereas chitosan and poly-L-lysine are cationic. This allows the electrostatic interaction of the two polymers to form a coat. It was found that during exposure to gastric and bile solutions those alginate microcapsules coated with chitosan demonstrated the best protection of the encapsulated cells. One possible pitfall of relying on electrostatic interactions is that alginate's degree of ionization will decrease during the reduction of pH associated with gastric passage; therefore it is important to understand the behaviour of these systems at the pHs associated with the GI tract. Such interactions could be studied using a variety of approaches, such as confocal microscopy using labelled polymers, turbidimetry, or the measurement of polymer dissociation by an appropriate quantification technique (*e.g.* HPLC). The interaction of alginate and chitosan with pH has been studied by turbidimetry,[18] revealing a bell-shaped curve between pH 1.8 and 10. This shows that over this pH range there are interactions occurring between alginate and chitosan, resulting in the formation of a turbid solution. However, at pHs below this, interactions may start to reduce and any polyelectrolyte complexes may dissociate, causing the removal of coat. This alginate–chitosan system has been adapted by Cook et al.[25] to introduce multilayer coatings to the outside of alginate microcapsules. Whereas the work by Krasaekoopt et al.[24] used a single layer of chitosan on the surface of the alginate, the repeated exposure of the microcapsule's surface to chitosan and alginate alternately allows the build-up of a thicker multilayer coat. This so-called layer-by-layer (LbL) technique allows for the simple modification of microcapsules, with a degree of control over the material properties by using a different numbers of coats on the surface of the alginate matrix. The ability to form these LbL coats was investigated by a surface plasmon resonance technique, allowing confirmation of interactions and study of their stability at GI pHs. It was found that by increasing the number of chitosan layers to three (which includes two coats of alginate) gave a stepwise improvement in the survival of a probiotic *Bifidobacterium* strain in simulated gastric solution. This technique is particularly attractive as a method for future research because of its simplicity, and the ease at which novel materials may be produced. One downside of the approach is that the exposure of alginate microcapsules to the coating solutions resulted in a slow loss of calcium ions from the centre of the alginate matrix. This resulted in swelling and loss of hardness as a result of a reduced crosslinking density, which may be an issue in further experimentation.[25] It should be noted that this crosslinking could be regained by the exposure of the microcapsules to calcium chloride solution.

An alternative to the coating of alginate microcapsules with polymers is to use proteins. The coating of these materials with proteins is attractive because of the high safety profile, cheapness and variety of food-grade proteins. For instance, Gbassi et al.[26] coated alginate microcapsules produced by extrusion and external gelation with whey protein. Whey protein is actually a mixture of globular proteins, namely β-lactoglobulin, α-lactalbumin, serum albumin and various immunoglobulins,[27] and was shown by the authors to be resistant to gastric fluids. The coating was achieved simply by dipping the alginate microcapsules into a solution of whey protein; association was presumably achieved by electrostatic interaction (though this was not stated). Once the alginate microcapsules had been coated with whey protein, they were exposed to a simulated GI transit and numbers of viable cells enumerated (Figure 5.6). This showed the coating of alginate microcapsules with whey protein to be an effective method of boosting the survival of three different strains of Lactobacillus plantarum during exposure to low pH, boosting survival over 2 h by over 5 log(CFU g^{-1}).

This study is particularly interesting because it opens the door to a variety of coatings consisting of other proteins, offering novel functional materials. The application of LbL coats utilizing proteins is also a possibility due to the ability of proteins to carry charge either side of their isoelectric point. Much work has been conducted on the production of protein–polymer multilayer films,[28] and this would appear to be a potentially effective avenue to explore. Additionally, the LbL coating of colloids has been achieved,[29] adding to the plausibility of this approach.

As mentioned earlier, an alternative approach to alteration of the alginate hydrogel is the incorporation of additional excipients into the gel network. This could either be achieved by using a colloidal dispersion or a polymer to

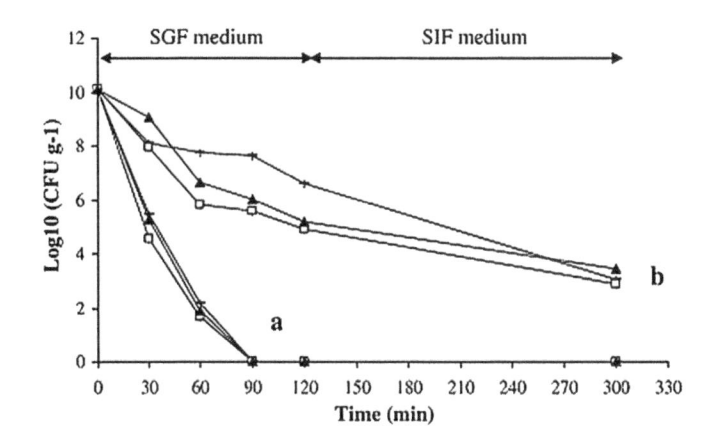

Figure 5.6 Survival of three Lactobacillus plantarum strains (strain numbers: – 299v, ▲ A159, □ 800) in simulated gastric fluid (SGF) and simulated intestinal fluid (SIF), encapsulated in (a) uncoated or (b) whey-protein coated alginate microcapsules. From ref. 26, with permission from Elsevier.

form a semi-interpenetrating polymer network. This 'blending' technique was used to effect by Sandoval-Castilla et al.[30] to produce a range of alginate–pectin microcapsules, of varying polymer ratios, containing *Lactobacillus casei*. Pectin is an interesting choice of polymer as, in its low methoxylated form, exhibits the same ability to form ionic crosslinks with calcium as alginate does. It was found that the exposure of these microcapsules to simulated gastric and bile salt conditions gave a reduced cell death as the concentration of pectin within the microcapsules increased. It should be noted that the alginate concentration was kept constant at 0.5% (w/v) and the pectin concentration increased from 1–3% (w/v). Thus, the microcapsules containing the highest ratio of pectin to alginate also contained the highest overall concentration of polysaccharide. This finding is also consistent with the hypothesis that cell survival in acid is improved by buffering within the hydrogel as an increase in overall polymer concentration led to an improved survival of cells. The alginate–pectin microcapsules were also effective at protecting cells during storage in yogurt, with the best survival seen at the highest pectin concentrations. Kim et al.[31] produced uniform 75 μm alginate–xanthan microcapsules using an air-atomizing device and external gelation. Though a control of alginate-only microcapsules was not present, these alginate–xanthan hybrid systems were very effective in protecting *Lactobacillus acidophilus* cells in simulated gastric conditions, despite their small size. This study gives hope that the modification of alginate hydrogels may be an effective method of counteracting the size-associated loss of effectiveness seen when trying to protect cells from acid.

5.6 The Future of Alginate as an Immobilization Matrix for Probiotics

Hopefully it has been demonstrated over the course of this chapter that the microencapsulation of probiotic bacteria into ionic alginate hydrogels is an effective method of improving the survival of these cells in low pH environments. These materials have application as a straightforward nutraceutical supplement, but also as a means of improving the shelf life of probiotic foods. Future challenges in this field arise from the downsizing of these materials and associated loss of protective effect, adverse effects on the textural properties of foods and effective human trials of these products. The problems associated with the loss of protective effect when downsizing alginate microcapsules could be counteracted by the use of coatings or additional excipients discussed previously. The production of smaller, yet still effective microcapsules should then allow for a product with improved sensory properties due to an improvement in texture. The need for effective human trials is the most challenging of these points. In an ideal world, the efficacy of an administration would be measured by its ability to provoke a response in humans. So, ideally the efficacy of a microencapsulated probiotic would be measured against its ability to alleviate a specific condition, such as ulcerative colitis. The few available human trials[21,32] measure numbers of viable cells found in the faeces, and so instead give an idea of the survival of the bacteria during GI transit.

There is also a need for the scale-up of these operations to be considered, as both the emulsion and extrusion methods used are batch processes. If this process could be made continuous, scale-up would be much more achievable and the mass production of these materials possible. A further trend is the application of these microencapsulation systems to a wider variety of food products, such as juices,[33] cheese,[34] and even more unusually, chocolate soufflé[35] and sausages.[36] So, to conclude, the future of these materials relies on the imagination of individual researchers taking these alginate microcapsules as a promising template for the future production of exciting, more functional, materials.

References

1. I. Metchnikov, *The prolongation of life*, G. P. Putnam, New York, 1908.
2. M. Shirota, Lactic acid bacteria, *Br. Pat.*, 1 167 196, 1969.
3. M. Boirivant and W. Strober, *Curr. Opin. Gastroenterol.*, 2007, **23**, 679.
4. H. Grasdalen, B. Larsen and O. Smidsrød, *Carbohydr. Res.*, 1979, **68**, 23.
5. H. Grasdalen, B. Larsen and O. Smisrød, *Carbohydr. Res.*, 1981, **89**, 179.
6. A. Martinsen, G. Skjåk-Bræk and O. Smidsrød, *Biotechnol. Bioeng.*, 1989, **33**, 79.
7. B. Thu, P. Bruheim, T. Espevik, O. Smidsrød, P. SoonShiong and G. Skjåk-Bræk, *Biomaterials,* 1996, **17**, 1031.
8. B. Rehm, *Alginates: Biology and Applications*, Springer, New York, 2009.
9. Ý. A. Mørch, I. Donati, B. L. Strand and G. Skjåk-Bræk, *Biomacromolecules,* 2006, 7, 1471.
10. K. I. Draget, G. Skjåk-Bræk and O. Smidsrød, *Carbohydr. Polym.*, 1994, **25**, 31.
11. I.-L. Andresen, O. Skipnes, O. Smidsrød, K. Østgaard and P. Hemmer, *ACS Symp. Ser.,* 1977, **48**, 361–381.
12. L. S. C. Wan, P. W. S. Heng and L. W. Chan, *Int. J. Pharm.*, 1994, **103**, 267.
13. D. Quong, R. Neufeld, G. Skjåk-Bræk and D. Poncelet, *Biotechnol. Bioeng.*, 1998, **57**, 438.
14. K. Ingar Draget, K. Østgaard and O. Smidsrød, *Carbohydr. Polym.*, 1990, **14**, 159.
15. J. C. Kolars, M. D. Levitt, M. Aouji and D. A. Savaiano, *N. Engl. J. Med.*, 1984, **310**, 1.
16. V. Chandramouli, K. Kailasapathy, P. Peiris and M. Jones, *J. Microbiol. Methods,* 2004, **56**, 27.
17. K.-Y. Lee and T.-R. Heo, *Appl. Environ. Microbiol.*, 2000, **66**, 869.
18. M. T. Cook, G. Tzortzis, D. Charalampopoulos and V. V. Khutoryanskiy, *Biomacromolecules,* 2011, **12**, 2834.

19. M. T. Cook, T. Saratoon, G. Tzortzis, A. Edwards, D. Charalampopoulos and V. V. Khutoryanskiy, *Biomacromolecules,* 2013, **14**, 387.
20. S. Graff, S. Hussain, J. C. Chaumeil and C. Charrueaul, *Pharm. Res.,* 2008, **25**, 1290.
21. J. H. Cui, Q. R. Cao and B. J. Lee, *Drug Delivery,* 2007, **14**, 265.
22. K. Kailasapathy, *Food Sci. Technol.,* 2006, **39**, 1221.
23. L. T. Hansen, P. M. Allan-Wojtas, Y. L. Jin and A. T. Paulson, *Food Microbiol.,* 2002, **19**, 35.
24. W. Krasaekoopt, B. Bhandari and H. Deeth, *Int. Dairy J.,* 2004, **14**, 737.
25. M. T. Cook, G. Tzortzis, V. V. Khutoryanskiy and D. Charalampopoulos, *J. Mater. Chem. B,* 2013, **1**, 52.
26. G. K. Gbassi, T. Vandamme, S. Ennahar and E. Marchioni, *Int. J. Food Microbiol.,* 2009, **129**, 103.
27. A. Haug, A. Hostmark and O. M. Harstad, *Lipids Health Dis.,* 2007, **6**, 1.
28. F. Caruso, K. Niikura, D. N. Furlong and Y. Okahata, *Langmuir,* 1997, **13**, 3427.
29. F. Caruso and H. Möhwald, *J. Am. Chem. Soc.,* 1999, **121**, 6039.
30. O. Sandoval-Castilla, C. Lobato-Calleros, H. S. García-Galindo, J. Alvarez-Ramírez and E. J. Vernon-Carter, *Food Res. Int.,* 2010, **43**, 111.
31. S.-J. Kim, S. Y. Cho, S. H. Kim, O.-J. Song, I. I. S. Shin, D. S. Cha and H. J. Park, *Food Sci. Technol.,* 2008, **41**, 493.
32. M. Del Piano, S. Carmagnola, S. Andorno, M. Pagliarulo, R. Tari, L. Mogna, G. P. Strozzi, F. Sforza and L. Capurso, *J. Clin. Gastroenterol.,* 2010, **44**, S42.
33. S. Nualkaekul, D. Lenton, M. T. Cook, V. V. Khutoryanskiy and D. Charalampopoulos, *Carbohydr. Polym.,* 2012, **90**, 1281.
34. H. Mirzaei, H. Pourjafar and A. Homayouni, *Food Chem.,* 2012, **132**, 1966.
35. C. Malmo, A. La Storia and G. Mauriello, *Food Bioprocess Technol.,* 2013, **6**, 795.
36. P. Muthukumarasamy and R. A. Holley, *Int. J. Food Microbiol.,* 2006, **111**, 164.

CHAPTER 6

Enzyme-Responsive Hydrogels for Biomedical Applications

YOUSEF M. ABUL-HAIJA AND REIN V. ULIJN*

Department of Pure and Applied Chemistry/WestCHEM, University of Strathclyde, 295 Cathedral Street, Glasgow G1 1XL, UK
*E-mail: rein.ulijn@strath.ac.uk

6.1 Introduction

Hydrogels are critically important components in biology as both the intracellular cytoskeleton and extracellular matrix (ECM) are gel-phase materials. Biological gels are highly responsive to changes in their environment, with enzymatic processes involved in adaption and reorganization of gel structures. This adaption and reorganization is key to processes such as differentiation and cell division. Given these observations, it is clear why an increasing number of researchers are focusing on the production of synthetic mimics of these biological hydrogels that may find applications in the measuring and directing of biological processes. In addition to the interest in enzyme-responsive systems in mimicking biological matrices, enzymatic processes are also useful in fabrication (enzyme-assisted assembly) of gels with precisely defined properties.[1] This chapter focuses on recent developments in these areas as well as the application of enzyme-responsive hydrogels in biomedicine, with a focus on supramolecular gels based on low molecular weight gelators (LMWG).

RSC Soft Matter No. 2
Hydrogels in Cell-Based Therapies
Edited by Che J. Connon and Ian W. Hamley
© The Royal Society of Chemistry 2014
Published by the Royal Society of Chemistry, www.rsc.org

6.1.1 Polymeric and Self-Assembling Hydrogels

Synthetic hydrogels have been considered in biomedical applications from 1960 when Wichterle and Lim demonstrated biocompatible hydrogels composed of cross-linked poly(hydroxyethylmethacrylic)acid [poly(HEMA)] that were later used in contact lenses.[2] Naturally derived hydrogels (alginates) were reported in the 1980s for encapsulation of pancreatic cells,[3] followed by the use of shark collagen as burn dressings.[4] Clearly, both synthetic and natural materials can be used effectively, with well-documented advantages and disadvantages for each of these. Hybrid systems, incorporating aspects of both are increasingly considered.[5] Early reports focused on the structural properties of hydrogels, which can match those of natural tissues. This focus has shifted to production of materials that do not just structurally, but also chemically, mimic aspects of biological gels, for example, by inclusion of short peptides, sugars or other biomolecules, to encourage 'active' interaction with biological systems. More recently, controlled degradation through enzyme action has become an additional feature of designed biomaterials.[1] The use of enzymes in the fabrication of materials has also become increasingly popular due to their selectivity and synthetic capability under mild conditions.[1]

Two types of synthetic gels can be distinguished, depending on the chemical nature of their networks (Figure 6.1). Polymeric hydrogels are covalently cross-linked networks that can swell due to the absorption and ability to trap water.[2] Supramolecular hydrogels are associated with reversible, non-covalent molecular interactions (hydrogen bonding, π-stacking, electrostatic and hydrophobic interactions, van der Waals forces) between self-assembling molecules (known as hydrogelators) to form nanofibres. These nanofibres then entangle into 3D networks that, depending on their surface chemistry, are able to trap water and form hydrogels. The design of polymeric hydrogels for biomedical applications is informed by substantial knowledge and literature on polymer synthesis and functionalization strategies. This is now a relatively mature field with many excellent contributions reported.[5] For supramolecular systems, the design rules are less well established and still emerging, with natural systems providing guidance and inspiration for their synthesis.[6]

6.1.2 Use of Enzymes in Fabrication of Next-Generation Biomaterials

One area where functionalized hydrogels have shown enormous potential is in gel-phase biomaterials for cell culture[6–9] and biosensing[10] platforms. In particular, there have been major developments in the application of hydrogels as instructive matrices for stem cell growth. The emphasis has been on the inclusion of biochemical signals, usually consisting of matrix protein specific peptidic motifs, such as the well-known fibronectin derived arginine–glycine–aspartic acid (RGD) motif that encourages cellular adhesion. A number of breakthroughs in recent years have shown that stem cell growth and differentiation, in addition to biochemical signals, are highly

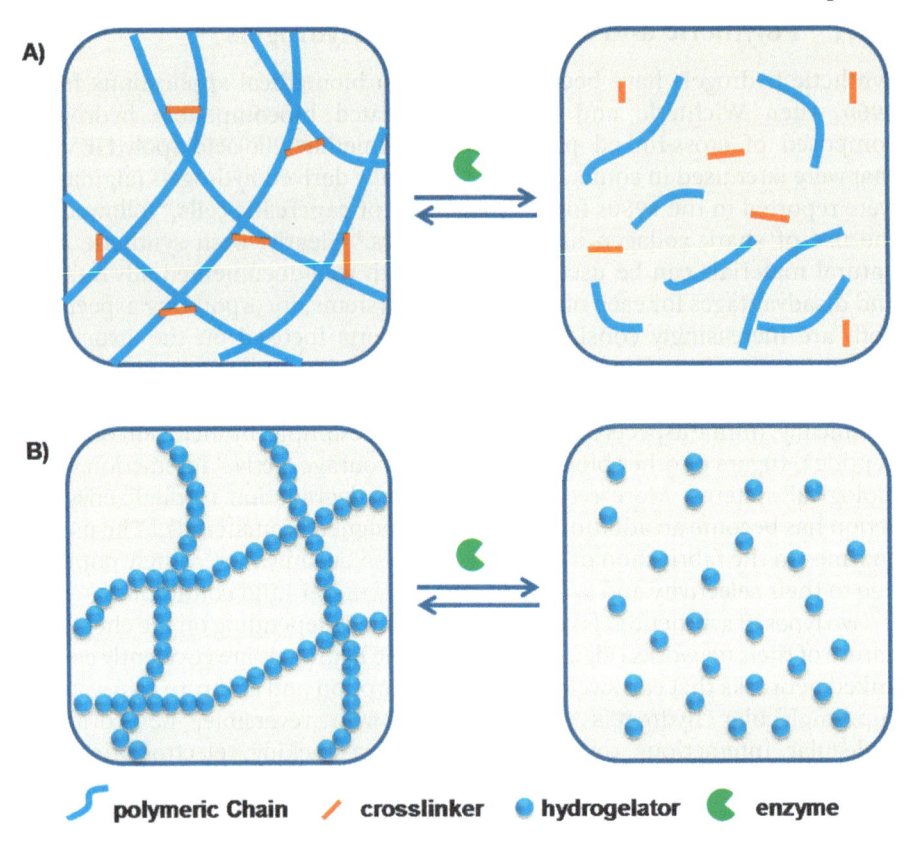

Figure 6.1 Schematic presentation of enzyme-responsive materials based on (A) polymeric and (B) supramolecular hydrogels. Enzymatic processes may be exploited in both the degradation and the controlled assembly of hydrogel materials.

sensitive to physical stimuli presented by their immediate environment.[11,12] Specifically, mechanical[13] (*i.e.* gel stiffness) and structural/topographical factors[14] of the cell-contacting matrix play crucial roles that have, in some cases, been shown to be more powerful than soluble biochemical signals. These approaches pave the way towards hydrogels that allow for the control of cell fate. Producing gels with specific requirements on physical properties and chemical composition will require new synthetic protocols that enable control of stiffness, gel network structure and chemical functionalization. It will be demonstrated that enzymatic fabrication methods provide useful tools to achieve this.[1,15]

6.1.3 Use of Enzymes as 'Stimuli' in Smart Materials

Since the early days of hydrogel research, efforts have been devoted towards the development of stimuli-responsive hydrogels as they potentially enable external control of cell encapsulation or release of actives. Stimuli-responsive, smart or intelligent hydrogels are those gels which may display property or functionality changes in response to variations in the external environment. Typically, these changes in their surroundings involve solvent polarity, temperature, pH, supply of electric field, light, *etc.*[16–19] More generally, materials based on stimuli-responsive technologies are increasingly attracting attention due to their potential applications in everyday life, offering improvements in many technologies.[1,20–24] There are excellent reviews on this topic that discuss the design, advantages and challenges of stimuli-responsive hydrogels.[18,25–28] In cellular environments, most stimuli-responsive mechanisms take place under the control of enzymes.[29] Compared with physical or conventional chemical stimuli (*e.g.* pH, temperature, ionic strength, ligand–receptor interactions), enzymatic regulation of materials properties shows much promise because it enables responsiveness to biology's own signals, which are highly selective and involve catalytic amplification to enable fast response times.[30,31] Examples have been reported on both polymeric[32,33] and supramolecular hydrogels in this context.

Controlled degradation of matrices is a particularly important focus area for enzyme-responsive materials. Hydrogels may be designed to render them degradable, which may involve breaking of chemical crosslinks (polymeric hydrogels) or controlled disassembly (supramolecular hydrogels) (Figure 6.1). Enzymatic degradation is selective and can in principle be tightly regulated, and is therefore highly attractive, as will be discussed below.

6.2 Biocatalytic Assembly of Supramolecular Hydrogels

Dynamic processes in biological systems are commonly controlled by spatially confined molecular mechanisms such as catalysis and molecular recognition. This is in contrast to traditional laboratory-based approaches to controlling supramolecular synthesis, which usually involves changing one or more of the environmental conditions such as pH,[34] temperature,[35] solvent polarity[36] and/or ionic strength.[37] In the last decade or so, the use of enzymes to direct supramolecular assembly has become more popular.[1,38–42] Biocatalysis is increasingly of interest in this context because (1) it allows responsive assembly under constant, physiological conditions;[43] (2) it allows exploitation of biocatalytic reactions which are specifically associated with certain cell types or diseased states;[30,44] (3) catalysis inherently involves molecular amplification (turnover numbers of 10^3–10^7 are common), which may give rise to fast response times; (4) it provides new tools for bottom-up nanofabrication by taking advantage of the ability to spatially and kinetically control the self-assembly process;[45] (5) thermodynamically controlled systems provide routes towards discovery of peptide-based nanostructures by exploiting reversible exchange of amino acid sequences in dynamic peptide libraries.[46–48]

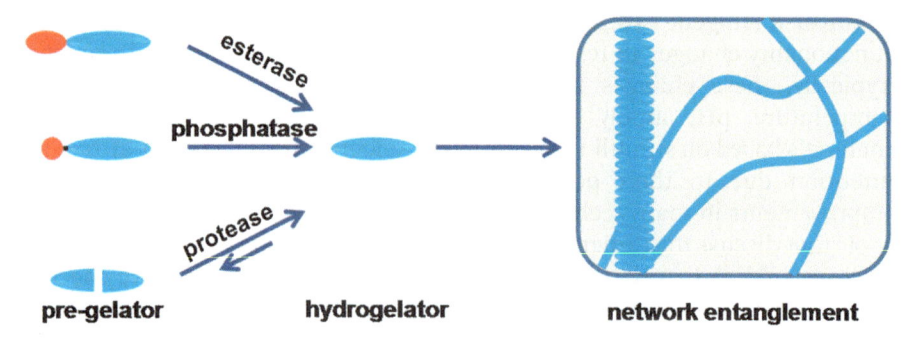

Figure 6.2 Schematic illustration of enzyme-assisted self-assembly. The enzyme action results in the formation the hydrogelators which are able to self-assemble to form supramolecular structures and then entangle to form a network.

Biocatalytic self-assembly involves the formation of gelators by the action of enzymes (either through hydrolysis or condensation of precursors), which is followed by the self-assembly of these molecules to form supramolecular structures. These assemblies in turn entangle to form 1D, 2D or 3D nanostructures (Figure 6.2) through non-covalent interactions such as π–π interactions, hydrogen bonding and electrostatic interactions.[49,50]

6.2.1. Peptide-Based Hydrogels

The majority of biocatalytic self-assembly systems reported to date are based on peptidic building blocks, so a brief introduction to this area is appropriate. Supramolecular hydrogels based on peptidic building blocks are of particular interest because of (1) their rich chemistry in non-covalent (hydrogen bonding, electrostatic, π-stacking, hydrophobic) interactions; (2) ease of synthesis and (3) biological compatibility. As peptides are biology's expression language, there is potential for them to be used as instructive biomaterials, *i.e.* containing peptide sequences found in the natural ECM.

There are 20 gene-encoded amino acids commonly found in nature. Depending on the nature of their side chains, non-covalent interactions can be incorporated at specific locations and the propensity for self-assembly thus varied. Considerable efforts have been made towards elucidating design rules for peptide-based supramolecular materials.[51–53] The design rules are either derived by copying nature (α-helix, β-sheet)[54,55] or are entirely new designs that exploit peptide derivatives such as aliphatic[56,57] or aromatic peptide amphiphiles.[21,39,58–60] While systems based on naturally occurring sequences typically use oligopeptides of 10 or more residues, the novel designs allow for the use of a minimalistic approach where much simpler and shorter peptides (as short as dipeptides) may be used. Aromatic peptide amphiphiles have been the basis of many biocatalytically triggered gels and are the focus of this chapter.

6.2.2 Biocatalytic Peptide Self-Assembly for Biomaterials Fabrication

The key point in enzyme-assisted self-assembly approaches is using an enzymatic bond-making/breaking reaction to regulate the balance between hydrophobicity and hydrophilicity to convert soluble precursor molecules into self-assembling hydrogelators, or *vice versa*. This can be achieved by two main methods: catalytic formation of a covalent bond to link two non-assembling components together to form a gelator, or catalytic cleavage of a covalent bond to transform a non-assembling precursor to a self-assembly building block. The first approach is achieved by catalysing the condensation reaction between two amino acid derivatives or peptide fragments. The second approach involves removing a functional group through a hydrolysis reaction, which might affect the molecular packing ability due to its bulky size or by causing electrostatic repulsion between monomeric units.[61]

Most commonly, phosphatases, esterases and proteases have been exploited as triggers for self-assembly of peptide derivatives (Figure 6.2). Enzymatic (de-)phosphorylation is used in biological systems to modify the structural features and biological activity of proteins. Self-assembly can be controlled using this method, driven by changes in electrostatic interactions as a result of the addition or removal of negatively charged phosphate groups. For example, Xu and Yang presented the first examples, where 9-fluorenylmethoxycarbonyl (Fmoc)–tyrosine modified with a phosphate group (Fmoc–Yp) dissolves in phosphate buffer solution. The electrostatic repulsion between phosphate groups of neighbouring molecules is known to prevent molecular self-assembly. Upon the addition of alkaline phosphatase enzyme to the peptide solution, Fmoc–Yp will be converted into Fmoc–Y, a more hydrophobic compound which is a hydrogelator that is able to self-assemble to form fibres.[62] We demonstrated that by extending the amino acid to a phenylalanyl-tyrosine (FY) dipeptide, thus enabling β-sheet interactions to contribute to the assembly, enzymatic dephosphorylation of Fmoc–FYp gives rise to a transformation of micelles to chiral, unidirectional fibres.[63] By exchanging phenylalanine (F) with other amino acids with different side chain properties, it becomes possible to form a range of different supramolecular structures[64-66] (*i.e.* different morphologies) which will be discussed further in the next section.

In order to provide a further step in understanding the mechanism of biocatalytic self-assembly, we have investigated the mechanism and kinetics of the phosphatase triggered self-assembly of Fmoc–Yp, as a model system.[67] We studied separately the biocatalytic conversion, changes in supramolecular interactions and chirality, nanostructure formation and gelation at varying enzyme concentrations giving rise to new mechanistic insights into the multi-stage self-assembly process. Furthermore, we observed a remarkable enhancement of catalytic activity during the early stages of the self-assembly process, providing evidence for enhancement of enzymatic activation by the supramolecular structures formed.

Esterases may be used to catalyse ester hydrolysis of ester-terminated peptides to form self-assembling peptides. We demonstrated that methyl esters of Fmoc–dipeptides could be converted into gelators by subtilisin, a common protease with esterase activity, resulting in formation of networks of nanotubes and fibres, depending on the peptide sequence.[45] Similarly, proteases can hydrolyse an amide bond to cleave a group associated with blocking.[68] Koda *et al.* have produced a hydrogelating species by hydrolysis of an amide bond, removing a fragment of the peptide chain which hinders self-assembly. The reaction was catalysed by matrix metalloproteinase-7 (MMP-7), an enzyme associated with degradation of ECM components by cells.[69] Additionally, β-lactamases have also been used to catalyse the formation of hydrogelators in the context of antibiotic-resistant bacteria that are known to express this type of enzyme. Here, the precursors were designed with a specific target for β-lactamase (*i.e.* a β-lactam ring), with the resulting cleaved molecule acting as a hydrogelator, with molecular self-assembly resulting in a nanofibrillar network structure.[70] The use of condensation reactions (*i.e.* the direct reversal of hydrolysis of amides) to produce gelators is a special case and will be discussed in the next section.

6.2.2.1 Morphology Control

Controlling the morphology of self-assembling peptide nanostructures continues to be an important challenge. As outlined in Section 6.1.2, fibre morphology (nanotopography) is an important variable in the design of supramolecular biomaterials, as it can impact dramatically on cellular response. Considerable research efforts have been devoted to develop the design rules for peptide building blocks to achieve morphological control which depends on two main principles: self-assembly pathway and peptide sequence.

6.2.2.1.1 Morphology Control by Pathway: Thermodynamic *vs.* Kinetic

It is increasingly recognized that the chosen self-assembly pathway dictates the structure of the final gel-phase material.[45,71] Biocatalytic self-assembly may proceed under kinetic control or thermodynamic control, depending on the reversibility of the enzyme reaction. Kinetic control gives rise to pathway-dependent properties, while thermodynamic control gives rise to reversible/self-healing materials that represent a free energy minimum stage. Both approaches are of interest in different contexts.

Biocatalytic Assembly Under Kinetic Control. Due to the restricted molecular dynamics in the gel phase, it is possible to trap molecular assemblies that do not represent the thermodynamically preferred structure. As a consequence, biocatalytic self-assembly may be used to form diverse materials from a single

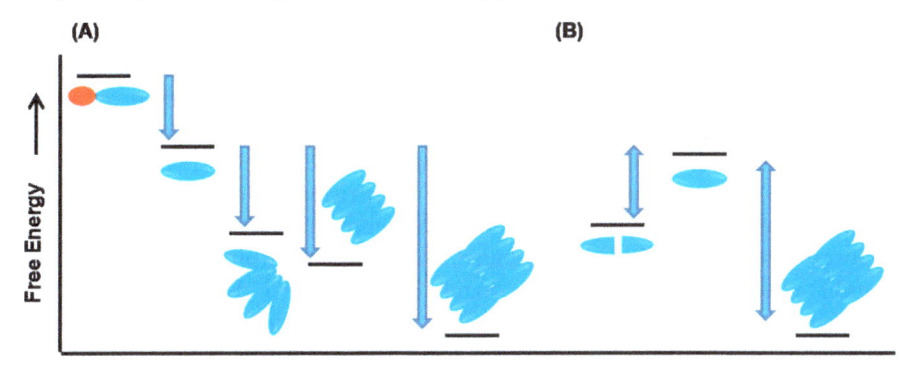

Figure 6.3 Free energy diagram of (A) kinetically controlled enzymatic self-assembly, where both the enzymatic reaction and the self-assembly processes are thermodynamically favoured. (B) Thermodynamically controlled enzymatic self-assembly, where the enzymatic reaction is thermodynamically unfavoured while the self-assembly process is energetically favoured.

gelator structure, regulation by the kinetics (enzyme concentration) of the reaction. Kinetic control is achieved when both the enzymatic reaction and the self-assembly process are independently favoured as illustrated in Figure 6.3, as observed for phosphatase[72] and subtilisin[73] responsive systems that operate by ester or phosphate ester hydrolysis, which readily proceeds in aqueous conditions. The rate of self-assembly in this case has substantial effect on the final structure, which is locked in the gel state. So, different supramolecular structures can be obtained by controlling the rate, which can be done by changing the amount of the catalyst, with stiffer gels generally resulting at higher rates of catalysis.[74] Upon gelation, systems become fixed in this arrangement, a process referred to as 'kinetic locking', as it is energetically unfavourable to reorganize the self-assembled monomers in combination with the extremely slow molecular dynamics associated with the gel phase. Consequently, the structures observed may not represent the global thermodynamic minimum.[73,75] Theoretically, by using a low concentration of enzyme molecules to control assembly, it would be possible to access the thermodynamically favoured structure to achieve thermodynamic control in these systems.

Using a kinetically controlled system can be advantageous. It is possible to control the properties of the hydrogels produced and allow tailoring materials for specific purposes while keeping the overall gelator concentration and chemical composition the same. For example, it has been shown that the mechanical properties of the hydrogels (*i.e.* stiffness) have been directly linked to the enzyme concentration. Higher enzyme concentration, leading to faster gel formation, could result in stiffer gels when directly compared with similar systems with lower enzyme concentration.[76,77]

Biocatalytic Assembly Under Thermodynamic Control. An enzymatically triggered system is said to operate under thermodynamic control when the enzymatic reaction producing self-assembling molecules is fully reversible. This is achieved when the enzymatic reaction is in itself energetically unfavourable (*e.g.* amide condensation, instead of hydrolysis) but is enabled by the favourable (small) free energy contribution of molecular self-assembly of the reaction product. This was demonstrated by the formation of a self-assembling peptide produced from non-assembling precursors, *via* amide condensation.[45] This approach gives rise to dynamic and fully reversible assemblies, which in turn, allow for defect correction and molecular reorganization, ultimately achieving a reproducible homogenous assembly of a structure representing the thermodynamic minimum.

The use of a thermodynamically driven approach introduces a means to directly compare the relative stabilities of each molecular building block and non-assembling precursor. Interestingly, when mixtures of precursors are supplied, these components would compete and finally self-select the most thermodynamically stable structure from dynamic mixtures. This type of system, where building blocks are continuously substituted until equilibrium is reached, is known as a dynamic combinatorial library (DCL).[78,79] This approach has been used in the discovery of stable supramolecular assemblies from mixtures. Due to the large possible peptide sequences that can be synthesized, the DCL approach is interesting for the identification of supramolecular peptide interactions.

Utilizing the enzymatic DCL approach, we screened a range of dipeptide sequences in Fmoc–dipeptide methyl ester gelators.[47] Peptide-based nanostructures have been investigated for the reaction of Fmoc–S and Fmoc–T with the amino acid methyl esters (X–OMe where X denotes leucine (L), phenylalanine (F), tyrosine (Y), valine (V), glycine (G) and alanine (A)) in presence of thermolysin to produce Fmoc–dipeptide esters, Fmoc–XY–OMe. It was found that the Fmoc–SF–OMe forms stable 2D sheet-like structures. Thus, the most stable self-assembled structures can be identified from a mixture of several components. This opens up the potential of exploiting the versatility of peptides for the discovery of functional nanostructures for various applications.

The use of enzyme-driven DCL approach was taken one step further with an unprecedented discovery of functional (rather than just structural) gels from component mixtures.[80] As shown in Figure 6.4, the DCL was generated by mixing Nap–Y donor, various amino acid amide nucleophiles (X–NH$_2$ where X denotes F, L, V, Y, A and G) and thermolysin in the absence and presence of dansyl-β-alanine acceptor (DA). In the absence of the acceptor, reversed-phase HPLC results showed that only a single component (Nap–YF–NH$_2$) was preferentially produced. Remarkably, similar experiments in the presence of DA (1 : 1 donor/acceptor ratio) showed that the formation of the major component YF was significantly amplified to 82% (instead of 52%), while the abundance of the second preferred compound, YL was reduced to 8% (instead of 23%). Furthermore, transmission electron microscopy showed the presence of entangled nanofibres of up to several micrometres in

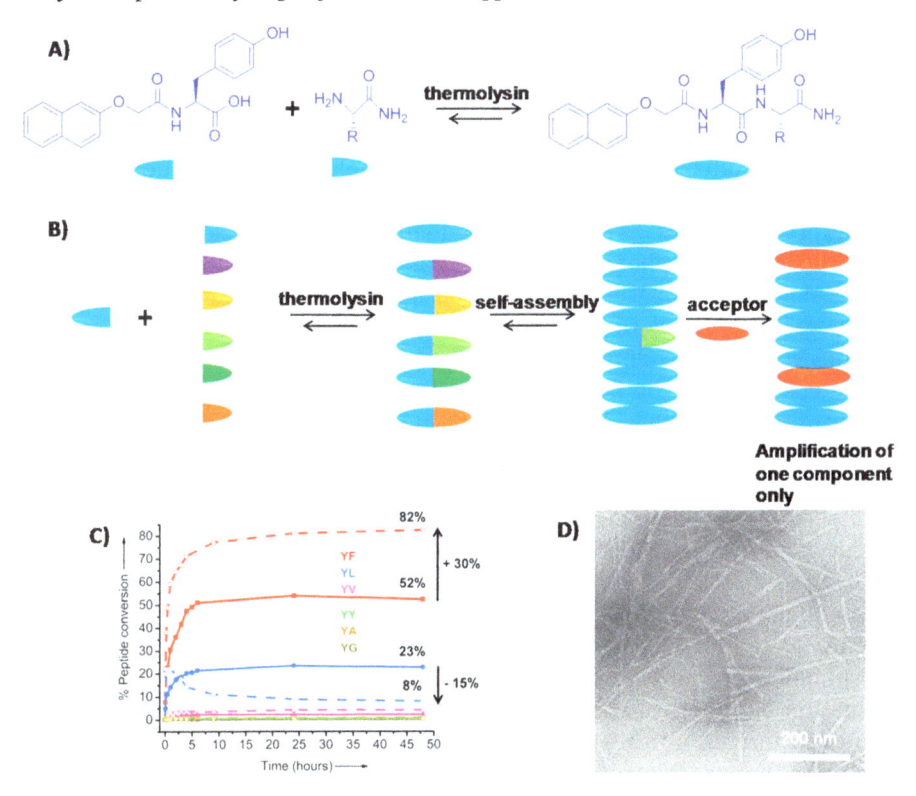

Enzyme-Responsive Hydrogels for Biomedical Applications 121

Figure 6.4 (A) Thermolysin-catalysed condensation reaction. (B) Schematic presentation of peptide library where the most stable product is preferentially amplified by acceptor molecule. (C) Time course of the percentage conversion in DCL system by HPLC as measured in the absence (solid traces) and presence (dashed traces) of DA library. (D) TEM image of presence of the acceptor molecules in DCL system. Reproduced from ref. 80.

length (Figure 6.4D). The presence of both donors and acceptors in the nanofibres resulted in an efficient energy-transfer observation between naphthalene donors and dansyl acceptors. Thus, the presence of suitable donors and acceptors in DCL greatly enhances its potential ability not just for self-selection but also for the amplification of most stable energy-transfer nanostructures at the expense of less stable self-assembling nanostructures. These energy-transfer nanostructures may pave the way towards development of gels for interfacing electronics with biology, as relevant for interfacing materials with neurons.

6.2.2.1.2 Morphology Control by Sequence

The second, perhaps more conventional, method to gain control over morphology is by changing the peptidic sequence. For aromatic peptide amphiphiles, it is known that self-assembly can be controlled by the combination of weak interactions. For these systems, directional control of self-assembly can be achieved through the aromatic π–π stacking. For example, it has been found that protecting the N-terminus of short peptides with an aromatic group (Fmoc, naphthalene, carbobenzyloxy) may enhance the formation of stable hydrogels. Fmoc–dipeptide gelators were first described in 1995 by Vegners et al.[81] It has been found that systematic variation of the side chains of the amino acids (i.e. peptide sequence) has a remarkable effect on the formed structures (spheres, fibres, tubes and sheets)[60,82] which is believed to have a significant impact on materials design for different purposes.

Our group studied the sequence–structure relationship in an enzyme-responsive aromatic dipeptides hydrogels under thermodynamic control, which ensures that the final self-assembled systems represent a thermodynamic minimum, rather than a trapped state. Systems studied were based on Fmoc-protected dipeptide methyl esters produced using thermolysin. We found that by varying the peptidic sequence, different morphologies could be accessed,[83] as shown in Figure 6.5A and B. This system remarkably led to the discovery of new, extended 2D self-assembled structures.[84]

Using a kinetically controlled reaction, we reported on subtilisin-catalysed formation of a range of peptide gelators with polar peptide functional groups (Figure 6.5C and D). The results demonstrated dramatic differences in morphology for four closely related Fmoc–dipeptide amphiphiles, including formation of spherical structures from Fmoc-YQ, while the closely related Fmoc-YN formed fibres. We demonstrated that the molecular packing abilities and hence the supramolecular structure are directly affected by changing the amino acid in position 2.[85] Recently, we produced the hydrogelators from phosphatase-responsive precursors, confirming that all peptide derivatives resulted in nanostructures with similar morphologies, independent of the enzymatic route used to produce them.[85]

It is clear that the combination of enzymatic reactions and peptide design provides access to a range of different structures. Using a combination of spectroscopy, microscopy techniques and molecular dynamics simulations, progress is being made with the elucidation of the 'design rules' for these systems. Ultimately, it should be possible to control the nanoscale topography and stiffness quite precisely, and in an on-demand fashion which will provide useful for future biomaterials design. In addition to structure, there is a need to control the chemical functionality of these systems, which will be described next.

6.2.2.2 Control of Functionality

It was previously found that Fmoc–dipeptide gelators can create hydrogel scaffolds that are compatible with certain cell types but produced fibres that were too hydrophobic for other cells to grow on.[86] Co-assembly of the Fmoc–dipeptide fibres with an Fmoc-protected amino acid or short peptide (surfactant-like peptide derivative) provides a handle for introducing chemical functionality, giving rise to more hydrophilic fibre surfaces that may

Enzyme-Responsive Hydrogels for Biomedical Applications 123

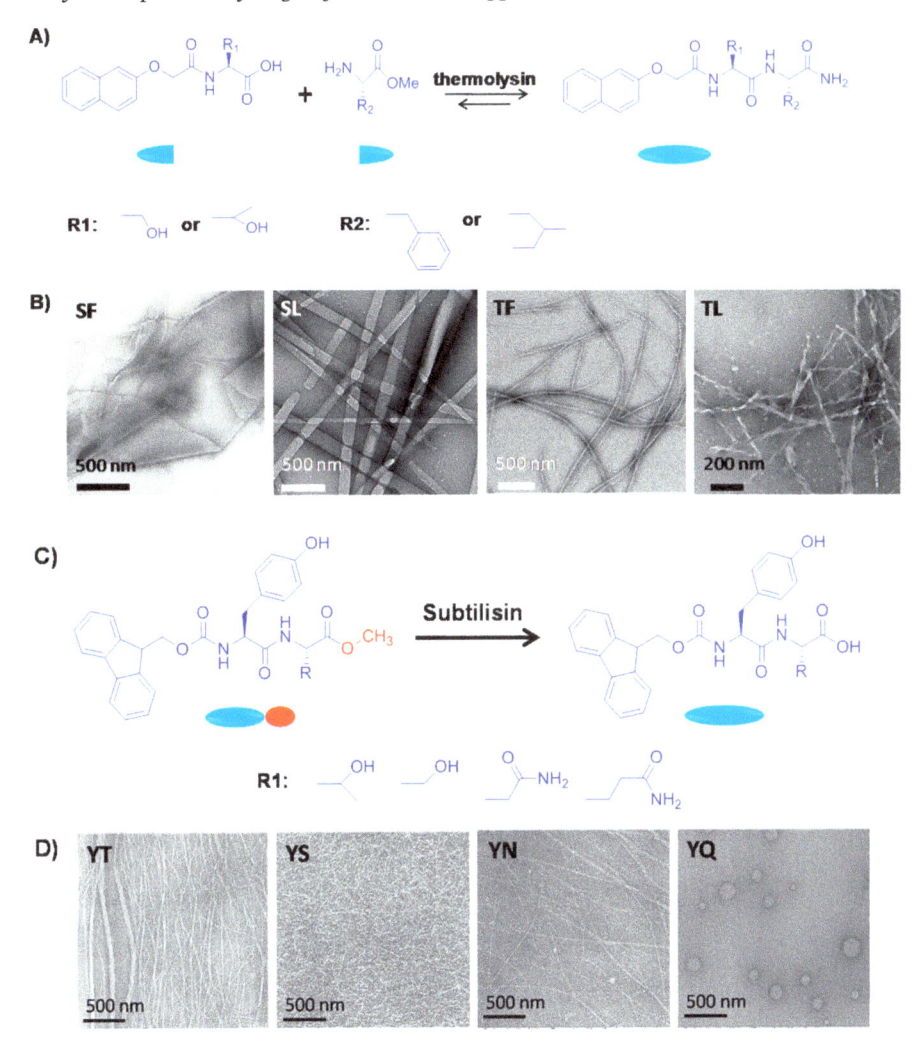

Figure 6.5 Controlling the morphology of enzyme-responsive hydrogels by changing the peptidic sequence. (A) Chemical structures of precursors and hydrogelators. (B) TEM images of Fmoc-protected dipeptides methyl esters produced by thermolysin. Reproduced from ref. 83. (C) Chemical structures of hydrogels produced by subtilisin-triggered hydrolysis of Fmoc–dipeptide esters. (D) TEM images of self-assembled structures. Reproduced from ref. 85.

contain functional peptides into the hydrogels. For example, co-assembly of Fmoc–diphenylalanine with Fmoc–serine gave rise to fibres that present the more favourable functionality presented by a serine derivative (hydroxyl and carboxylic acid), thus greatly enhancing cytocompatibility. In a recent study, we demonstrated a facile supramolecular approach for the formation of functionalized nanofibres by combining the advantages of biocatalytic

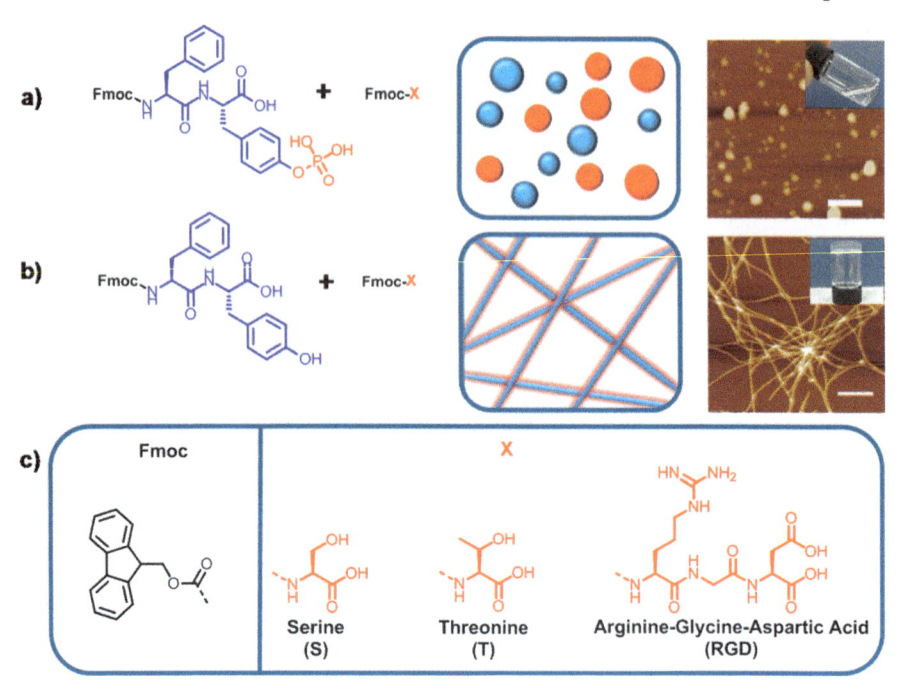

Figure 6.6 Functional nanofibres formed as a result of enzyme triggered co-assembly. (a) Both Fmoc–FY*p* and Fmoc–X [X = serine (S), threonine (T) or arginine–glycine–aspartic acid (RGD)] form spherical aggregates (blue and red spheres) in solution at physiological conditions which self-assemble into (b) surfactant coated-fibrous gels when triggered by alkaline phosphatase. (c) Chemical structures of Fmoc moiety and X peptide moieties. Scale bar = 2 µm. Reproduced from ref. 87.

self-assembly and surfactant/gelator co-assembly.[87] This is achieved by enzymatically triggered reconfiguration of free-flowing micellar aggregates of pre-gelator (Fmoc–FY*p*) and functional surfactants Fmoc–X (X = S, T or RGD, where RGD is the well-known cell adhesion motif[88]) to form nanofibres that become coated with the surfactants (Figure 6.6). This results in the formation of fibres that display the functionality at the surface. Furthermore, by varying enzyme concentration, the gel stiffness and supramolecular organization of building blocks can be varied.

6.3 Biomedical Applications

It is clear from the above that enzyme-responsive hydrogels, based on peptide self-assembly, are gaining increasing interest as biomaterials. The main application areas that are envisaged are cell culture, drug delivery, biosensing and, increasingly, the ability to control and direct cell fate in a dynamic manner using systems that operate intracellularly. In this section, we provide

Enzyme-Responsive Hydrogels for Biomedical Applications 125

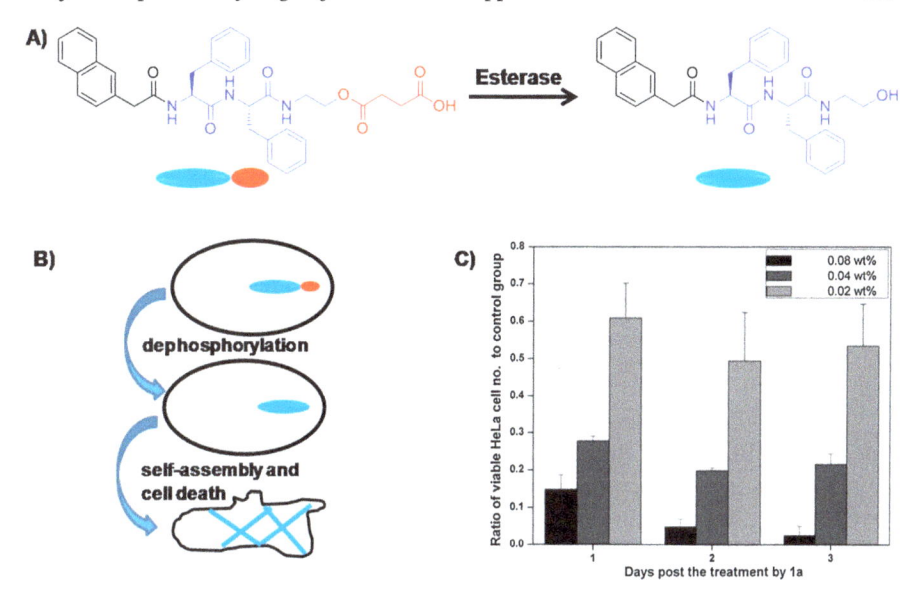

Figure 6.7 (A) Chemical structures of intracellular cleavage of an esterase responsive precursor by an endogenous esterase to a hydrogelator. (B) Corresponding schematic representation for the formation of supramolecular assembly within the cells. (C) MTT assay results of HeLa cells treated with different concentrations (0.2%, 0.4% and 0.8%) of the precursor. Reproduced from ref. 30.

details on some recent highlights that pave the way towards potential biomedical applications of enzyme-responsive hydrogels.

6.3.1 Controlling and Directing Cell Fate

As molecular self-assembly and catalysis are two of the key molecular mechanisms that drive dynamic processes in cells, it makes sense to harness these mechanisms and produce systems that allow researchers to interfere with these cellular processes. The first example that utilized enzyme-responsive hydrogels intracellularly was published by Xu and his co-workers. They designed a peptide-based precursor which could be triggered by an esterase enzyme to form self-assembled nanostructures to induce gelation inside the cell. When exposed to cells, the precursor enters into the cell by diffusion and undergoes hydrolysis by an endogenous esterase enzyme to form a hydrogelator which self-assembles into nanostructures (Figure 6.7).[30,44] This gelation induced an abrupt change in the viscosity of the cell cytoplasm and caused cell death. At a certain concentration, the majority of cells derived from human cervical cancer tumours died within 3 days, whereas fibroblast cells remained alive under the same conditions. Moreover, it also found that the formation kinetics of the intracellular nanofibres was specific to different

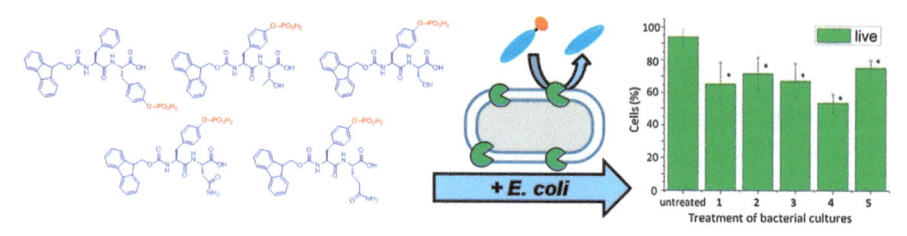

Figure 6.8 Chemical structures of alkaline phosphatase-responsive amphiphilic precursors and the percentage of viable (live) cells in bacterial cultures after treatment with the phosphorylated precursors of self-assembling aromatic peptide amphiphiles. Reproduced from ref. 89.

types of cells because of different esterase expression levels,[44] which may open up new ways to treat tumour with high selectivity (*e.g.* tumour cells usually have higher esterase expression levels than normal cells).

Xu and co-workers exploited enzymatic hydrogelation to control bacterial cells. They designed a phosphatase-responsive precursor that remains in solution in the absence of enzyme. *E. coli* is known to over-express phosphatase enzymes in the periplasmic space. By using the over-expressed alkaline phosphatase in *E. coli* bacteria, this precursor is converted to the corresponding hydrogelator inside the bacterial cell. The subsequent intracellular hydrogelation inhibited the bacterial growth.[30] Experiments showed that intracellular enzymatic formation of supramolecular nanostructures can be used for the development of antimicrobial biomaterials.

In order to investigate whether nanostructural morphology could impact on *E. coli* cell fate,[30] we used a range of different aromatic peptide amphiphiles which are known to form different supramolecular structures. We investigated whether the enzymatic formation of nanostructures within the periplasmic space (where phosphatases reside) may provide a powerful means to achieve antimicrobial activity.[89] First, we studied the self-assembly of a number of designed phosphatase-responsive precursors based on Fmoc-protected amphiphiles; namely, FY*p*, Y*p*T, Y*p*S, Y*p*N and Y*p*Q. Then, we investigated whether the antimicrobial activity can be differentially controlled by the introduction of these aromatic peptide amphiphiles. While there was a substantial difference in uptake of the gelators by cells, no significant difference in the antimicrobial response was observed for different treatments, all showing modest antimicrobial activity (Figure 6.8).

6.3.2 Imaging and Biosensing

An approach was developed for fluorescent intracellular imaging by Xu and co-workers by using gelators with an incorporated fluorophore, which intensifies upon self-assembly.[90] After an enzymatic conversion inside the cell, the precursor turns to the corresponding hydrogelator, a more hydrophobic molecule that is able to self-assemble to form nanofibres. When the

Enzyme-Responsive Hydrogels for Biomedical Applications 127

precursors are outside cells or the concentration of hydrogelator is too low to form nanofibres, those precursors or hydrogelators diffuse, distribute homogeneously, fluoresce identically within each pixel and thus show little contrast. Once the concentration of hydrogelator reaches high enough to form nanofibres, these nanofibres have more fluorophores within each pixel than the rest of the solution, and the fluorophores within nanofibres are localized within the cell, therefore the nanofibres fluoresce more brightly and generate the contrast and in principle allow for the assessment of the locale of enzyme action within the cell (Figure 6.9). The approach is potentially very powerful in visualizing different enzyme activities (protease, phosphatase, esterase, *etc.*) within different cellular compartments.

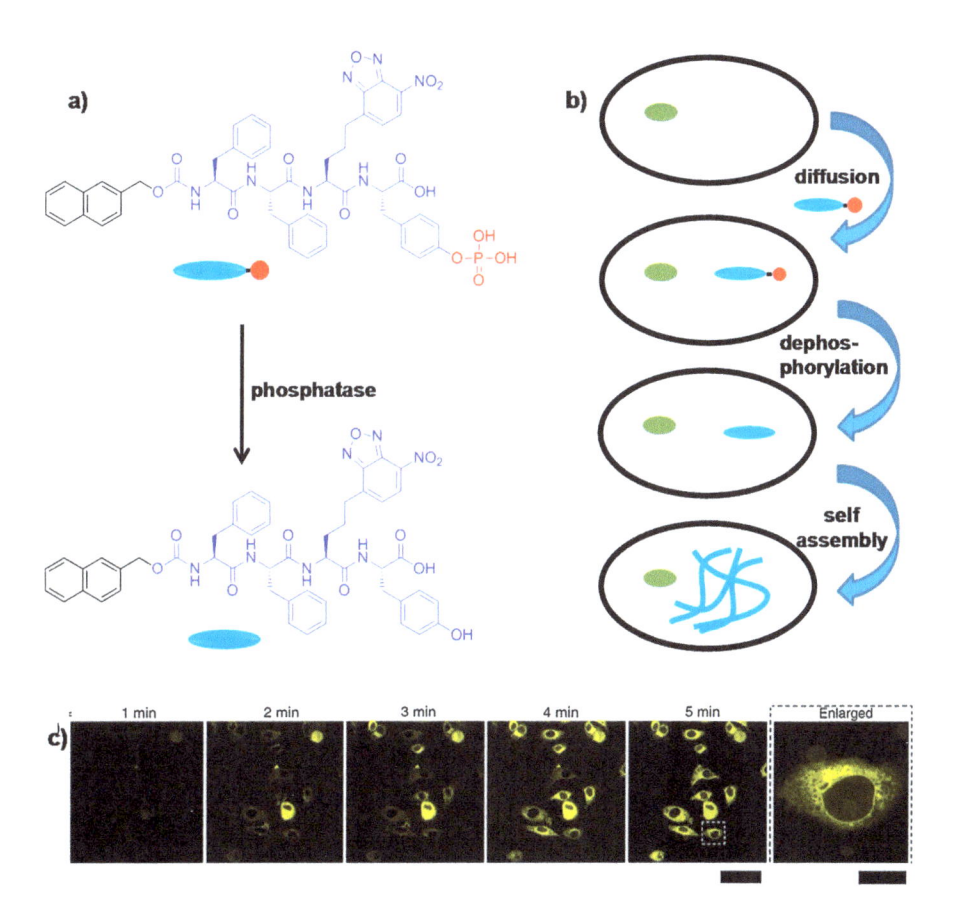

Figure 6.9 (a) Chemical structure of phosphatase-responsive compound which forms a hydrogel under the action of enzyme. (b) Schematic representation of the precursor diffusion into the cell and subsequent dephosphorylation and self-assembly. (c) Time course of fluorescent confocal microscope images inside the HeLa cells. Scale bar = 50 μm for time course images and 10 μm for the enlarged image. Reproduced from ref. 90.

In recent work, we made a first step towards developing an optical sensor for enzyme activity by interfacing an enzyme-responsive hydrogel with a liquid crystal (LC) display. This work was inspired by reports by the Abbott group,[91] who showed that enzyme activity could be directly visualized using LCs. In proof-of-concept work, we made use of a two-stage process, involving Fmoc–TL–OMe produced by thermolysin-catalysed condensation. The OMe group can be subsequently hydrolysed by subtilisin, which results in disassembly. A dual-layer design was developed, where a phospholipid-loaded upper gel layer (Fmoc–TL–OMe) was separated from the LC display by a phospholipid-free lower Fmoc–TL–OMe layer. When subtilisin was applied, it was shown to digest both layers, releasing the phospholipids to give a gel-to-sol transition after several hours that liberated the phospholipid and produced a light-to-dark optical change in the LC display.[92] In future, it may be possible to use similar designs to achieve sensing of other biologically relevant events, including digestion of extracellular matrix by cell secreted enzymes.

6.3.3 Controlled Drug Release

An interesting supramolecular design approach towards a drug delivery application was reported by Gao *et al.* In this work, taxol (an anticancer drug) was modified to form nanofibres which were found to be able to entangle to form a supramolecular hydrogel. The drug was covalently attached to a precursor which bears a part cleavable by phosphatase.[93] Hydrolysis of phosphate group occurs upon exposure to alkaline phosphatase and hydrogel can be formed and subsequently taxol derivative release. This derivative showed similar activity to taxol in toxicology studies.

Most recently, Gao *et al.* exploited a tyrosinase enzyme to control a supramolecular disassembly process which was considered to be potentially useful for controlled drug release in exploiting elevated tyrosinase activity in malignant melanoma. Congo red (as a model drug) was incorporated within hydrogel matrix assembled from aromatic tetrapeptide methyl esters, Ac–YYYY–OMe and Ac–FYYY–OMe.[66] Upon treatment with tyrosinase, tyrosine residues were converted to quinone. This oxidation process results in the loss of π–π interactions between phenol rings and ultimately a gel-to-solution phase transition which in turn results in release of drug molecules. The incorporated drug could be released in a controllable manner by using different enzyme concentrations.

6.3.4 Cell Scaffolds and Tissue Engineering

A first cell culture example that makes use of biocatalytic self-assembly was published by Richard Williams' group (Figure 6.10).[94] They demonstrated the use of (immobilized) thermolysin-catalysed condensation of Fmoc–L and L_2 to form a self-assembling tripeptide derivative. It was demonstrated that laminin could be distributed throughout the network *via* non-covalent

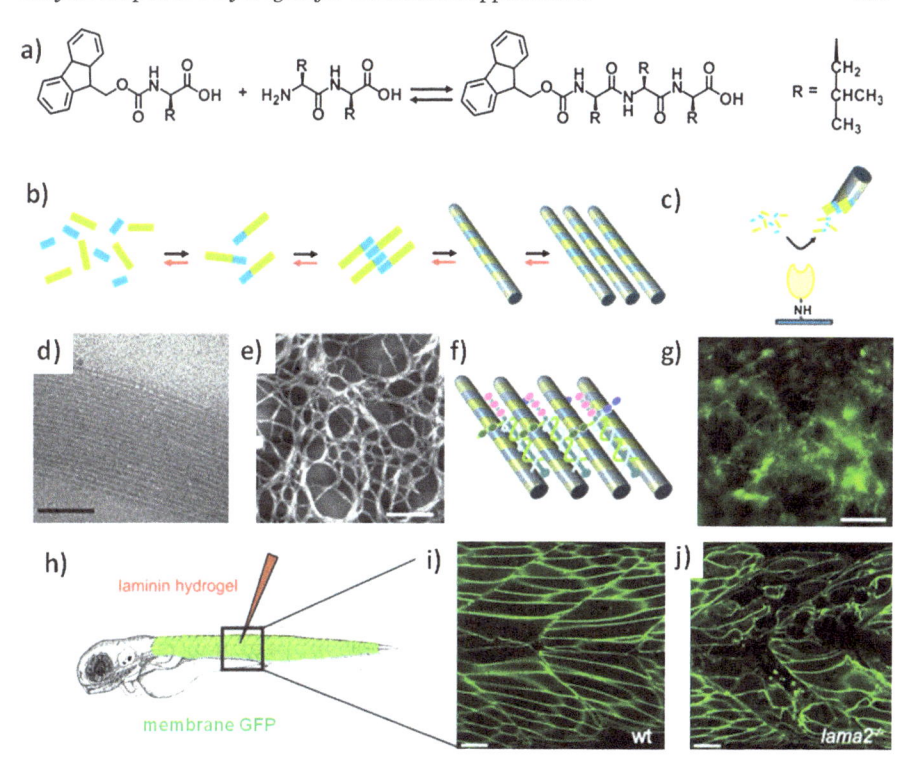

Enzyme-Responsive Hydrogels for Biomedical Applications 129

Figure 6.10 *In vivo* performance of hydrogel produced by biocatalytic peptide self-assembly. (a) Fmoc–amino acid reacts with a dipeptide in the presence of the enzyme, to yield the hydrogelator Fmoc–trileucine. (b) Formation of self-assembling fibres *via* thermolysin-catalysed condensation (c) using an immobilized enzyme. (d) TEM (scale bar 200 nm) and (e) AFM of gel fibres (scale bar 2 mm). (f) Complexation with laminin as verified by fluorescence microscopy (g, scale bar 2 mm). (h) Cartoon of the site of hydrogel injection in a 3-day-old laminin-deficient zebrafish. (i) Confocal microscope image of intact muscle in wild-type *wt* zebrafish (scale bar 25 mm). (j) Corresponding image of muscle from a laminin-deficient (lama2$^{-/-}$) dystrophic zebrafish (scale bar 25 mm). From ref. 94, with permission. Copyright Elsevier 2011.

interactions. The stability and suitability of the material for *in vivo* use was tested utilizing microinjection into a dystrophic zebrafish model organism. This organism lacks laminin as a result of a genetic mutation; instead, laminin is provided by the self-assembled gel system. It could be confirmed that the biomaterial remained *in situ* at the site of injection.

6.4 Conclusions and Outlook

There is an increasing interest in mimicking biological systems in an effort to enhance control over the bottom-up fabrication process. Events in biological systems are often controlled by spatially confined molecular mechanisms such as catalysis and molecular recognition. Therefore, enzymes are identified as a useful handle to dictate such nanofabrication process towards developing complex and highly selective next-generation biomaterials. Enzyme-assisted formation and dismantling of supramolecular structures have many properties such as self-assembly under constant conditions, spatiotemporal control of nucleation and structure growth, and control of mechanical properties (*i.e.* stiffness). Moreover, systems which assemble under thermodynamic control have particular features such as defect-correcting and component-selecting abilities. Researchers have devoted their attention towards designing materials for particular applications in cell culture, drug delivery, imaging, biosensing, and controlling cell fate.

Dynamic processes in biological systems are usually controlled by enzymes. Therefore, there are many attempts to overcome challenges in understanding these processes at the molecular level through enzymatic controlled self-assembly approach. Key challenges are (1) to control nucleation and structure growth, (2) to access structures that represent non-equilibrium assemblies and (3) to produce asymmetric, dynamic and multi-component structures with desired functionalities.

It is becoming increasingly obvious that studying systems based on enzyme responsiveness will become of great interest not only for the formation of supramolecular structures but also for studying the ability of these structures to recognize, adapt, correct, and interact in complex ways and give rise to evolving behaviours. This can be demonstrated, for example, in the field of specific disease control where the development of molecular materials that respond to a disease-specific molecular event is highly desired.

Acknowledgments

The authors would like to acknowledge the financial support by FP7 Marie Curie Actions of the European Commission, *via* the initial training network ReAd (Contract No. 289723).

References

1. M. Zelzer, S. J. Todd, A. R. Hirst, T. O. McDonald and R. V. Ulijn, *Biomater. Sci.,* 2013, **1**, 11–39.
2. O. Wichterle and D. Lim, *Nature,* 1960, **185**, 117–118.
3. F. Lim and A. M. Sun, *Science,* 1980, **210**, 908–910.
4. I. V. Yannas, E. Lee, D. P. Orgill, E. M. Skrabut and G. F. Murphy, *Proc. Natl. Acad. Sci. U. S. A.,* 1989, **86**, 933–937.
5. B. Ratner, A. S. Hoffman, F. Schoen and J. E. Lemons, *Biomaterials Science: Introduction to Materials in Medicine*, Elsevier Academic Press, San Diego, CA, 2nd edn, 2004, pp. 162–164.
6. G. A. Silva, C. Czeisler, K. L. Niece, E. Beniash, D. A. Harrington, J. A. Kessler and S. I. Stupp, *Science,* 2004, **303**, 1352–1355.

7. P. Y. Dankers, M. C. Harmsen, L. A. Brouwer, M. J. Van Luyn and E. Meijer, *Nat. Mater.*, 2005, **4**, 568–574.
8. E. F. Banwell, E. S. Abelardo, D. J. Adams, M. A. Birchall, A. Corrigan, A. M. Donald, M. Kirkland, L. C. Serpell, M. F. Butler and D. N. Woolfson, *Nat. Mater.*, 2009, **8**, 596–600.
9. F. Gelain, D. Bottai, A. Vescovi and S. Zhang, *PLoS One*, 2006, **1**, e119.
10. S. Kiyonaka, K. Sada, I. Yoshimura, S. Shinkai, N. Kato and I. Hamachi, *Nat. Mater.*, 2003, **3**, 58–64.
11. D. E. Discher, D. J. Mooney and P. W. Zandstra, *Science*, 2009, **324**, 1673–1677.
12. M. Lutolf and J. Hubbell, *Nat. Biotechnol.*, 2005, **23**, 47–55.
13. A. J. Engler, S. Sen, H. L. Sweeney and D. E. Discher, *Cell*, 2006, **126**, 677–689.
14. M. J. Dalby, N. Gadegaard, R. Tare, A. Andar, M. O. Riehle, P. Herzyk, C. D. Wilkinson and R. O. Oreffo, *Nat. Mater.*, 2007, **6**, 997–1003.
15. M. Ehrbar, S. C. Rizzi, R. G. Schoenmakers, B. San Miguel, J. A. Hubbell, F. E. Weber and M. P. Lutolf, *Biomacromolecules*, 2007, **8**, 3000–3007.
16. B. D. Ratner, A. S. Hoffman, F. J. Schoen and J. E. Lemons, *Biomaterials Science: Introduction to Materials in Medicine*, Elsevier Academic Press, San Diego, CA, 2nd edn, 2004.
17. J. Kopecek, *Eur. J. Pharm. Sci.*, 2003, **20**, 1.
18. J. F. Mano, *Adv. Eng. Mater.*, 2008, **10**, 515–527.
19. D. Schmaljohann, *Adv. Drug Delivery Rev.*, 2006, **58**, 1655–1670.
20. S. Koutsopoulos, L. D. Unsworth, Y. Nagai and S. Zhang, *Proc. Natl. Acad. Sci. U. S. A.*, 2009, **106**, 4623–4628.
21. A. M. Smith, R. J. Williams, C. Tang, P. Coppo, R. F. Collins, M. L. Turner, A. Saiani and R. V. Ulijn, *Adv. Mater.*, 2008, **20**, 37–41.
22. L. Chen, K. Morris, A. Laybourn, D. Elias, M. R. Hicks, A. Rodger, L. Serpell and D. J. Adams, *Langmuir*, 2009, **26**, 5232–5242.
23. M. Zhou, A. M. Smith, A. K. Das, N. W. Hodson, R. F. Collins, R. V. Ulijn and J. E. Gough, *Biomaterials*, 2009, **30**, 2523–2530.
24. Y. C. Jung, H. Muramatsu, K. Fujisawa, J. H. Kim, T. Hayashi, Y. A. Kim, M. Endo, M. Terrones and M. S. Dresselhaus, *Small*, 2011, 7, 3292–3297.
25. C. de las Heras Alarcón, S. Pennadam and C. Alexander, *Chem. Soc. Rev.*, 2005, **34**, 276–285.
26. K. Soppimath, T. Aminabhavi, A. Dave, S. Kumbar and W. Rudzinski, *Drug Dev. Ind. Pharm.*, 2002, **28**, 957–974.
27. M. Prabaharan and J. F. Mano, *Macromol. Biosci.*, 2006, **6**, 991–1008.
28. R. Kulkarni and S. Biswanath, *J. Appl. Biomater. Biomech.*, 2007, **5**, 125.
29. C. Walsh, *Nature*, 2001, 226–231.
30. Z. Yang, G. Liang, Z. Guo and B. Xu, *Angew. Chem., Int. Ed.*, 2007, **46**, 8216–8219.
31. J. M. Spruell and C. J. Hawker, *Chem. Sci.*, 2011, **2**, 18–26.
32. J. L. West and J. A. Hubbell, *Macromolecules*, 1999, **32**, 241–244.
33. P. D. Thornton and A. Heise, *Chem. Commun.*, 2011, **47**, 3108–3110.

34. C. Tang, A. M. Smith, R. F. Collins, R. V. Ulijn and A. Saiani, *Langmuir,* 2009, **25**, 9447–9453.
35. H. Hong, Y. Mai, Y. Zhou, D. Yan and Y. Chen, *J. Polym. Sci., Part A: Polym. Chem.,* 2007, **46**, 668–681.
36. M. Reches and E. Gazit, *Science,* 2003, **300**, 625–627.
37. B. Ozbas, J. Kretsinger, K. Rajagopal, J. P. Schneider and D. J. Pochan, *Macromolecules,* 2004, **37**, 7331–7337.
38. Z. Yang, H. Gu, D. Fu, P. Gao, J. K. Lam and B. Xu, *Adv. Mater.,* 2004, **16**, 1440–1444.
39. Z. Yang, G. Liang and B. Xu, *Acc. Chem. Res.,* 2008, **41**, 315–326.
40. J. H. Collier and P. B. Messersmith, *Bioconjugate Chem.,* 2003, **14**, 748–755.
41. S. Winkler, D. Wilson and D. Kaplan, *Biochemistry,* 2000, **39**, 12739–12746.
42. S. Toledano, R. J. Williams, V. Jayawarna and R. V. Ulijn, *J. Am. Chem. Soc.,* 2006, **128**, 1070–1071.
43. B. Xu, *Langmuir,* 2009, **25**, 8375–8377.
44. Z. Yang, K. Xu, Z. Guo, Z. Guo and B. Xu, *Adv. Mater.,* 2007, **19**, 3152–3156.
45. A. R. Hirst, S. Roy, M. Arora, A. K. Das, N. Hodson, P. Murray, S. Marshall, N. Javid, J. Sefcik, J. Boekhoven, J. H. van Esch, S. Santabarbara, N. T. Hunt and R. V. Ulijn, *Nat. Chem.,* 2010, **2**, 1089–1094.
46. R. J. Williams, A. M. Smith, R. Collins, N. Hodson, A. K. Das and R. V. Ulijn, *Nat. Nanotechnol.,* 2008, **4**, 19–24.
47. A. K. Das, A. R. Hirst and R. V. Ulijn, *Faraday Discuss.,* 2009, **143**, 293–303.
48. J. W. Sadownik and R. V. Ulijn, *Chem. Commun.,* 2010, **46**, 3481–3483.
49. D. M. Ryan and B. L. Nilsson, *Polym. Chem.,* 2012, **3**, 18–33.
50. D. J. Adams and P. D. Topham, *Soft Matter,* 2010, **6**, 3707–3721.
51. A. R. Hirst, I. A. Coates, T. R. Boucheteau, J. F. Miravet, B. Escuder, V. Castelletto, I. W. Hamley and D. K. Smith, *J. Am. Chem. Soc.,* 2008, **130**, 9113–9121.
52. M. de Loos, B. L. Feringa and J. H. van Esch, *Eur. J. Org. Chem.,* 2005, **2005**, 3615–3631.
53. H. Kobayashi, A. Friggeri, K. Koumoto, M. Amaike, S. Shinkai and D. N. Reinhoudt, *Org. Lett.,* 2002, **4**, 1423–1426.
54. D. N. Woolfson and M. G. Ryadnov, *Curr. Opin. Chem. Biol.,* 2006, **10**, 559–567.
55. Y. Yanlian, K. Ulung, W. Xiumei, A. Horii, H. Yokoi and Z. Shuguang, *Nano Today,* 2009, **4**, 193–210.
56. L. A. Estroff and A. D. Hamilton, *Chem. Rev.,* 2004, **104**, 1201–1218.
57. R. V. Ulijn, *J. Mater. Chem.,* 2006, **16**, 2217–2225.
58. M. Reches and E. Gazit, *Nat. Nanotechnol.,* 2006, **1**, 195–200.
59. R. V. Ulijn and A. M. Smith, *Chem. Soc. Rev.,* 2008, **37**, 664–675.
60. D. J. Adams, M. F. Butler, W. J. Frith, M. Kirkland, L. Mullen and P. Sanderson, *Soft Matter,* 2009, **5**, 1856–1862.
61. M. Reches and E. Gazit, *Curr. Nanosci.,* 2006, **2**, 105–111.

62. Z. Yang and B. Xu, *Chem. Commun.,* 2004, 2424–2425.
63. J. W. Sadownik, J. Leckie and R. V. Ulijn, *Chem. Commun.,* 2011, **47**, 728–730.
64. H. Wang, C. Ren, Z. Song, L. Wang, X. Chen and Z. Yang, *Nanotechnology,* 2010, **21**, 225606.
65. Q. Wang, Z. Yang, Y. Gao, W. Ge, L. Wang and B. Xu, *Soft Matter,* 2008, **4**, 550–553.
66. J. Gao, W. Zheng, D. Kong and Z. Yang, *Soft Matter,* 2011, 7, 10443–10448.
67. K. Thornton, Y. M. Abul-Haija, N. Hodson and R. V. Ulijn, *Soft Matter,* 2013, **9**, 9430–9439.
68. S. Dos Santos, A. Chandravarkar, B. Mandal, R. Mimna, K. Murat, L. Saucede, P. Tella, G. Tuchscherer and M. Mutter, *J. Am. Chem. Soc.,* 2005, **127**, 11888–11889.
69. D. Koda, T. Maruyama, N. Minakuchi, K. Nakashima and M. Goto, *Chem. Commun.,* 2010, **46**, 979–981.
70. Z. Yang, P.-L. Ho, G. Liang, K. H. Chow, Q. Wang, Y. Cao, Z. Guo and B. Xu, *J. Am. Chem. Soc.,* 2007, **129**, 266–267.
71. H. Wang, Z. Yang and D. J. Adams, *Mater. Today,* 2012, **15**, 500–507.
72. Z. Yang, H. Gu, D. Fu, P. Gao, J. K. Lam and B. Xu, *Adv. Mater.,* 2004, **16**, 1440–1444.
73. A. R. Hirst, S. Roy, M. Arora, A. K. Das, N. Hodson, P. Murray, S. Marshall, N. Javid, J. Sefcik, J. Boekhoven, J. H. van Esch, S. Santabarbara, N. T. Hunt and R. V. Ulijn, *Nat. Chem.,* 2010, **2**, 1089–1094.
74. R. J. Williams, R. J. Mart and R. V. Ulijn, *Biopolymers,* 2010, **94**, 107–117.
75. S. Roy and R. V. Ulijn, *Advances in Polymer Science, in Enzymatic Polymerisation,* ed. A. R. A. Palmans and A. Heise, Springer, Berlin, 2010, vol. 237, pp. 127–143.
76. K. Thornton, A. M. Smith, C. L. R. Merry and R. V. Ulijn, *Biochem. Soc. Trans.,* 2009, **37**, 660–664.
77. Z. Yang, G. Liang and B. Xu, *Soft Matter,* 2007, **3**, 515–520.
78. P. T. Corbett, J. Leclaire, L. Vial, K. R. West, J.-L. Wietor, J. K. M. Sanders and S. Otto, *Chem. Rev.,* 2006, **106**, 3652–3711.
79. S. J. Rowan, S. J. Cantrill, G. R. L. Cousins, J. K. M. Sanders and J. F. Stoddart, *Angew. Chem., Int. Ed.,* 2002, **41**, 898–952.
80. S. K. M. Nalluri and R. V. Ulijn, *Chem. Sci.,* 2013, **4**, 3699–3705.
81. R. Vegners, I. Shestakova, I. Kalvinsh, R. M. Ezzell and P. A. Janmey, *J. Pept. Sci.,* 1995, **1**, 371–378.
82. Y. Zhang, H. Gu, Z. Yang and B. Xu, *J. Am. Chem. Soc.,* 2003, **125**, 13680–13681.
83. M. Hughes, P. W. J. M. Frederix, J. Raeburn, L. S. Birchall, J. Sadownik, F. C. Coomer, I.-H. Lin, E. J. Cussen, N. T. Hunt and T. Tuttle, *Soft Matter,* 2012, **8**, 5595–5602.
84. M. Hughes, H. Xu, P. W. J. M. Frederix, A. M. Smith, N. T. Hunt, T. Tuttle, I. A. Kinloch and R. V. Ulijn, *Soft Matter,* 2011, **7**, 10032–10038.
85. M. Hughes, L. S. Birchall, K. Zuberi, L. A. Aitken, S. Debnath, N. Javid and R. V. Ulijn, *Soft Matter,* 2012, **8**, 11565–11574.

86. V. Jayawarna, S. M. Richardson, A. R. Hirst, N. W. Hodson, A. Saiani, J. E. Gough and R. V. Ulijn, *Acta Biomater.*, 2009, **5**, 934–943.
87. Y. M. Abul-Haija, S. Roy, P. W. J. M. Frederix, N. Javid, V. Jayawarna and R. V. Ulijn, *Small*, 2013, DOI: 10.1002/smll.201301668.
88. E. Ruoslahti, *Annu. Rev. Cell Dev. Biol.*, 1996, **12**, 697–715.
89. M. Hughes, S. Debnath, C. W. Knapp and R. V. Ulijn, *Biomater. Sci.*, 2013, **1**, 1138–1142.
90. Y. Gao, J. Shi, D. Yuan and B. Xu, *Nat. Commun.*, 2012, **3**, 1033.
91. J. M. Brake, M. K. Daschner, Y.-Y. Luk and N. L. Abbott, *Science*, 2003, **302**, 2094–2097.
92. I. H. Lin, L. S. Birchall, N. Hodson, R. V. Ulijn and S. J. Webb, *Soft Matter*, 2013, **9**, 1188–1193.
93. Y. Gao, Y. Kuang, Z.-F. Guo, Z. Guo, I. J. Krauss and B. Xu, *J. Am. Chem. Soc.*, 2009, **131**, 13576–13577.
94. R. J. Williams, T. E. Hall, V. Glattauer, J. White, P. J. Pasic, A. B. Sorensen, L. Waddington, K. M. McLean, P. D. Currie and P. G. Hartley, *Biomaterials*, 2011, **32**, 5304–5310.

CHAPTER 7

Alginate Hydrogels for the 3D Culture and Therapeutic Delivery of Cells

BERNICE WRIGHT* AND CHE J. CONNON

School of Chemistry, Food and Pharmacy, University of Reading, Reading, RG6 6AD, UK
*E-mail: b.wright@ucl.ac.uk; c.j.connon@reading.ac.uk

7.1 Alginate Isolation and Gelation Chemistry

The biophysical properties and biocompatibility of alginate hydrogels are significantly dependent on their source, extraction methodology, polymer composition, and gelation chemistry. Alginates isolated from various brown algae (seaweed, *e.g. Ascophyllum nodosum*, *Laminaria* spp., *Lessonia nigrescens, Ecklonia maxima, Macrocystis pyrifera* and *Durvillaea antarctica*)[1,2] and certain bacterial species (*e.g. Pseudomonas* and *Azotobacter* genus)[3] are dissimilar with regard to β-D-mannuronic acid (M) and α-L-guluronic acid (G) content. Alginate gel mechanical strength, elasticity and swelling characteristics vary depending on the M:G ratio, length of MMM, GGG and MGM polymer blocks, and the percentage of these block structures. Multivalent cations (*e.g.* Ca^{2+}, Ba^{2+}, La^{3+}, Fe^{3+}, Zn^{2+}, Mg^{2+}, Sr^{2+}) used to crosslink carboxyl groups on M and G polymers to form alginate gels also determine mechanical strength. Furthermore, the extraction of alginate polysaccharides affects the viscosity of alginate gels, a property that strongly influences their biocompatibility and clinical application. These physical properties must be considered collectively to produce alginate hydrogels that are suitable for biomedical application.

RSC Soft Matter No. 2
Hydrogels in Cell-Based Therapies
Edited by Che J. Connon and Ian W. Hamley
© The Royal Society of Chemistry 2014
Published by the Royal Society of Chemistry, www.rsc.org

7.1.1 Extraction and Purification of Alginate Polysaccharides

Commercial alginates are extracted mainly from brown algae (*Phaeophyceae*) and telluric bacteria (acetylated alginate). Extraction and purification of alginate involves precipitation, quantification of uronic acids, complete and partial hydrolysis to separate MG, M and G blocks, spectroscopic analyses to confirm purification and determine structure, and rheology measurements to measure viscosity.

Methodology for extracting polysaccharides from algae involves the use of solvents that allow a series of precipitation steps before purification. A procedure developed by Calumpong *et al.*[4] is based on the method that is commonly used for extracting alginate from algae. This method involves soaking dried algae in 2% formaldehyde for 24 h at room temperature before washing samples in water and incubating them in 0.2 M HCl for a further 24 h. Algae samples are washed again with water before extraction for 3 h at 100 °C using 2% sodium carbonate. The soluble fraction is then collected by centrifugation and polysaccharides are precipitated using three volumes of 95% ethanol, they are washed with acetone, dried at 65 °C and dissolved in water. A final precipitation step is performed in ethanol and the sodium alginate is dried at 65 °C. The amount of alginate extracted can be measured by evaluating the levels of acidic D-guluronic acid present in the extracted alginate using colorimetric assays which distinguish neutral D-glucose sugars from acidic D-guluronic acid sugars. The uronic acid (M and G) content of alginate can be quantified by absorbance at 490 nm after reaction with *m*-hydroxydiphenyl.

The separation of MG, M and G blocks from purified alginate is achieved by partial hydrolysis.[5] This method can be performed in a series of relatively simple steps. Initially, the purified polysaccharide is heated at reflux with 3 M HCl for 20 min. After cooling, the suspension is centrifuged (3000 × g, 20 min) and the supernatant is neutralized (1 M NaOH) prior to supplementation with ethanol. At the end of this first stage, the precipitate (**MG block**) is collected by centrifugation (10 000 × g, 20 min), dissolved in water and freeze-dried. For the second stage, the insoluble fraction from the first centrifugation is heated at reflux with 0.3 M HCl for 2 h, centrifuged (10 000 × g, 20 min), neutralized (1 M NaOH), and the pH is adjusted to 2.85 with 1 M HCl, before a further neutralization step, and dialysis against water. The insoluble fraction when freeze-dried is the **M block**. The precipitate from the final centrifugation step is dissolved by neutralization, dialysed and freeze-dried to obtain the **G block**.

Methods for structural characterization of purified alginate generally include Fourier transform infrared (FT-IR) or proton nuclear magnetic resonance (^{1}H NMR) spectroscopy.[6-8] NMR is considered the most reliable method for determining the composition and block structure of alginate, but both techniques are applied to obtain confident measurements of purity. The average molecular weight and molecular weight distribution is determined by high-pressure size-exclusion chromatography (HPSEC) with online multi-angle laser light scattering (MALLS).[7,8] The principle of this method is to separate individual particles and then determine their absolute molar mass and average size by detecting how they scatter light. After analysis of the purity of extracted alginate polymers, rheological measurements of alginate are typically performed to assess the viscosity of the polysaccharide. One method for measuring alginate viscosity is by performing cone-plate geometry on a stress-controlled rheometer, where viscosity is measured as a function of shear rate.[6–8]

The methods for alginate extraction and purification are well established and they yield a material that is safe enough for wide-ranging use in the food industry as thickeners, emulsifying agents and components for the production of gels.[9] Therefore, biomedical use of alginate polysaccharides is dependent only on identifying modifications that can be made to this polymer to produce appropriate scaffolds for particular clinical applications or cell types.

7.1.2 Alginate Gelation Chemistry

Alginates are true block structures consisting of homopolymeric regions of (1,4) linked β-D-mannuronic acid (M) with a $^4C_1$ ring conformation and α-L-guluronic acid (G) with a $^1C_4$ ring conformation, interspersed with regions of alternating structure (MG blocks).[10] In solution, alginates are arranged as flexible coils. These polysaccharides are polyelectrolytes; therefore the selective binding of certain alkaline multivalent earth metals (*e.g.* Ca^{2+}, Ba^{2+}, La^{3+}, Fe^{3+}, Zn^{2+}, Mg^{2+}, Sr^{2+}) lead to their formation into a gel.[11,12] Upon interaction with divalent metal cations this material gels into an ordered structure (Figures 7.1 and 7.2).

The gelation process mediated by calcium ions (Ca^{2+}) involves cooperative binding of ions between aligned GG blocks of two alginate chains.[10] Calcium ions cause chain–chain associations which represent the junction zones responsible for gel formation. These zones are commonly known as the 'egg-box model'.[10,11] In this model, pairs of 2_1 helical chains are packed with the calcium ions located between them.[11] The gelation of alginate is controlled by the relative rate of diffusion of ions and polymer molecules into the gelling zone.[13,14] Calcium alginate gels are not, however, uniform, mechanically strong or easily manipulated into complex 3D structures. Attempts to control the gelation rate of these gels to generate more structurally uniform and mechanically stronger gels include the use of $CaCO_3$–D-glucono-δ-lactone (GDL) and $CaSO_4$–$CaCO_3$–GDL systems which slow down the gelation process.[15]

The use of monovalent cations (sodium (Na^+) and potassium (K^+)) together with calcium (Ca^{2+}) has also been investigated to examine the rate of calcium-mediated sol–gel transitions.[16] A previous study demonstrated that at apparent equilibrium and at low calcium concentration, sodium alginate gels displayed reduced elastic moduli compared to potassium alginate gels. Sol–gel transition kinetics and modulus at apparent equilibrium, increased

138　　Chapter 7

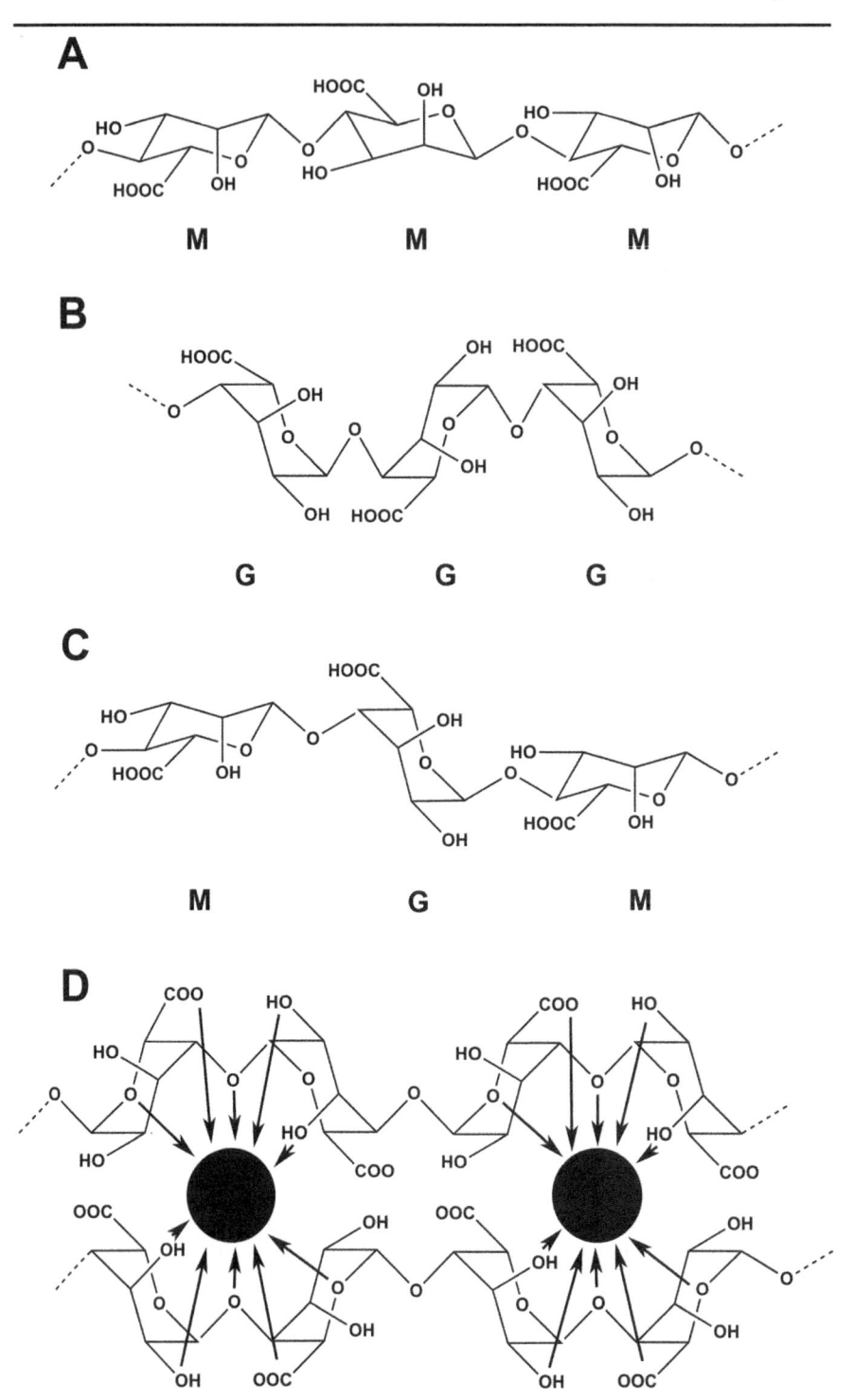

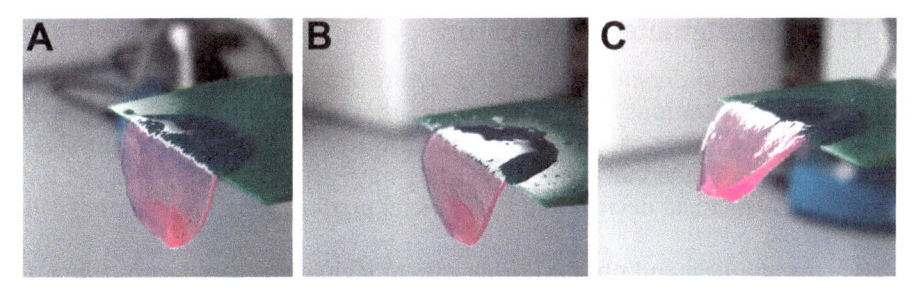

Figure 7.2 Alginate hydrogels. Alginate gels of various dimensions can be formed. The images show calcium alginate gel discs of approximately 19 mm in length and 1.5 mm in depth containing 1.2% (w/v) (A), 2.4% (w/v) (B) and 3.6% (w/v) (C) alginate polymer.

with increased content of guluronic acid residues in the alginate sample, and an increased amount of internally released Ca^{2+} (15–30 mM) resulted in equal elastic properties at apparent equilibrium. The study suggested that dissimilarities in ion exchange reactions between Ca^{2+}/Na^+ and Ca^{2+}/K^+ may be the underlying cause for these gelation kinetics.

Alginate gels are often chemically modified to create mechanically stable structures that can be loaded with cells and remain at sites of injury for the periods of time required to allow the repair of damaged tissue. The polymer block structures of alginates and the type of crosslinking cations used can influence the strength of gels.[17] The effect of different alginates and crosslinking ions on the stability and strength of alginate microcapsules was recently investigated.[17] The dimensional stability and gel strength increased for high-G alginate gelled with barium (Ba^{2+}) compared to samples gelled using calcium (Ca^{2+}). Barium decreased the size of alginate beads and decreased their permeability to immunoglobulin G. Gelation with strontium (Sr^{2+}) produced gels with characteristics intermediate between calcium and barium. High-M alginate beads gelled using barium and strontium were larger than beads of calcium alginate and the former gels were more permeable. Furthermore, calcium was shown to bind to G- and MG blocks, and barium to G- and M-blocks, and strontium bound only to G blocks.

Taken together, these published studies show that the process of forming alginate gels can give rise to numerous different permutations with variations in physical properties (elasticity, mechanical strength) that can be exploited to construct macrostructures potentially tailored to diverse biomedical applications.

Figure 7.1 The structure and gelation chemistry of alginate. Alginate comprises of β-D-mannuronic acid (M) and α-L-guluronic acid (G) polymer blocks [MMM, GGG and MGM], and multi-valent cations (*e.g.* Ca^{2+}, Ba^{2+}, La^{3+}, Fe^{3+}, Zn^{2+}, Mg^{2+}, Sr^{2+}: indicated by black circle) are used to crosslink carboxyl groups on M and G polymers to form alginate gels.

7.1.3 The Alginate Gel Structure

A fundamental understanding of the structure of alginate gels is important to perform modifications to this material that allow an effective clinical endpoint. A number of studies have investigated the viscoelastic, physicochemical and functional properties of alginate gels to explore avenues for their use as medical devices and bioartificial platforms for investigating cell function.

The mechanical and physical properties of alginate gels impact in a major way on cells. Therefore, these structures are often finely tuned to direct the function and phenotype of encapsulated cells. Alginate gels should be stable for the period of time that they are needed for cell delivery, but maintaining the mechanical strength of these substrates in physiological environments is difficult. The main cause of alginate polycation capsule breakage under physiological conditions is probably osmotic swelling of the alginate core, caused by the Donnan equilibrium set up by the negative charges of the carboxyl groups that are not involved in cooperative binding of counterions in the junction zones of chain networks.[18]

Reducing the swelling capacity of alginate gel microcapsules was previously demonstrated to stabilize these scaffolds.[18] Stable alginate polycation capsules were made by increasing the strength of the polyanion–polycation membrane, or by keeping a low-swelling gel network in the core using alginates rich in guluronic acid both in the core and in an outer coating. The polymer material was anisotropically distributed in the core; the concentration at the surface was higher than that in the centre of the capsule.

Studies have shown that high-G alginate polymers yield stronger, more ductile hydrogels than high-M alginates.[19] Another method for producing mechanically strong alginate gels is through the use of secondary polymers to reinforce the surface of these scaffolds. A recent study hypothesized that the molecular weight of the polymer used to reinforce alginate would be an important factor in the stability of this gel, particularly when the gel network was homogeneously reinforced with the polymer.[20] This hypothesis was investigated with alginate hydrogels crosslinked with calcium, and reinforced throughout the bulk of the gel with poly(ethyleneimine) (PEI) of varying molecular weights. Interactions between the two polymers became significant following gelation, leading to higher elastic moduli than gels with no PEI. The stable interactions between the alginate and PEI prevented alterations of the pore structure in the gels, and slowed the deterioration of gel properties.

The viscoelastic properties and permeability of alginate gels determine the efficacy of this substrate as a model system of soft biological tissues. Characterization of alginate viscoelastic properties was performed previously by measuring the elastic modulus of this material.[21] A report examined the time-dependent deformation of thin circular alginate gel membranes under a constant load. The elastic modulus as a function of time was determined by a large deformation theory based on Mooney–Rivlin elasticity, and a viscoelastic theory, Zener model, was applied to correlate the time-dependent deformation of the constructs with various gel concentrations.[22,23] By contrast, a recent report described the 'stretchability' of alginate gels and outlined the synthesis of gels that formed into ionically, and covalently crosslinked networks.[24] The novel gels could be stretched beyond 20 times their initial length; this was attributed to crack bridging by the network of covalent crosslinks, and hysteresis by unzipping the network of ionic crosslinks.

The permeability of alginate gels is important in circumstances when this scaffold is used for the encapsulation and delivery of cells, which require the mass transfer of nutrients and metabolic products.[25,26] A study described alginate microcapsules designed for the delivery of cells based on polyelectrolyte complexation, where the inner phase involved a blend of alginate and sodium cellulose sulfate (SCS).[27] The molar mass of the SCS was shown to influence transport properties of the scaffold. The ideal ratio, for mechanical and transport properties, of SCS to alginate was 55:45 (w/w).

These studies demonstrate that control of the structure and physical properties of alginate may be performed in an elegant manner, to introduce micro-scale features into this material that aid the maintenance of encapsulated cells.

7.2 Alginate Hydrogels as Cell Culture Scaffolds

Alginate hydrogels used for cell encapsulation and culture are on the threshold of widespread application in the clinical setting. They are already established as dermal bandages routinely used in the clinic for the treatment of burn injuries. Embryonic, mesenchymal, neural and cardiac stem cells, chondrocytes and corneal cells are amongst the more promising cell types amenable to therapeutic delivery with alginate gels as a support scaffold. Those cell types have been demonstrated to remain functional and promote tissue regeneration, whilst encapsulated in alginate hydrogels.

The early promise of stem cell therapy may be fulfilled through the application of alginate gels as delivery vehicles, as the structure of this material is compliant to modifications which mimic the physiological environment of cells. The development of bioartificial niches specific to a particular cell type is therefore possible with alginate. Those biomimetic systems may either be used as medical devices to immobilize cells in a functional state that is therapeutic, and/or as tools to aid understanding of the regulation of cells.

7.2.1 Regeneration of the Cornea Using Alginate-Encapsulated Corneal Cells

The cornea is the outermost transparent region of the eye, and is composed of a number of different cell types which maintain transparency and hydration and allow regeneration of the surface of this structure. Adult stem cells which replenish cells in the cornea reside in the palisades of Vogt[28,29] in the outer limbal region. Corneal transparency is due to a smooth epithelium with an underlying stromal layer consisting of uniformly spaced collagen fibres[30] and interconnected keratocytes which produce crystalline proteins,[30,31] as well as an endothelium that regulates corneal hydration.[32]

The use of alginate hydrogels for ocular therapy is not established. Alginate hydrogels as part of composite biomaterial systems were previously applied for ocular reconstruction. Alginate microspheres incorporated into collagen hydrogels were demonstrated as a viable composite construct for controlled drug delivery as well as human corneal epithelial cell growth.[33] Alginate membranes coated with chitosan and used as base matrices for limbal epithelial cell cultivation maintained the attachment, spreading and growth of these cells.[34] Another composite hydrogel containing sodium alginate dialdehyde and hydroxypropyl chitosan was successfully used for transplantation of corneal endothelial cells (CEC) onto Descemet's membrane; this construct demonstrated that encapsulated CEC remained viable and retained their normal morphology.[35]

The well-reported use of alginate gels for cell preservation[36–40] may allow this hydrogel to be developed into a transport/storage medical device applicable to corneal cell-based therapies. A recent study outlined promising results for the short-term storage and transport of limbal epithelial cells (LEC) in structurally-modified alginate discs.[36] A range of technologies that apply alginate hydrogels as a starting material may therefore be developed for corneal regeneration.

7.2.2 Harnessing the Therapeutic Potential of Embryonic Stem Cells Using Alginate Gels

Embryonic stem cells (ESC) are pluripotent stem cells derived from the inner cell masses of early-stage embryos (blastocysts), and are the most attractive alternative sources for cell replacement therapy due to their pluripotency and their capacity for indefinite expansion in culture.[41] A number of factors limit the therapeutic application of these cells, however. ESC require specialized culture conditions that are labour-intensive[42] and involve the use of potentially pathogenic xenogeneic sources of growth factors.[43,44] Moreover, ESC proliferation and differentiation are dependent on external signal autonomous regulators, *i.e.* a niche-like environment[45,46] that is not successfully reproduced in the 2D systems primarily used for the culture of these cells. Alginate hydrogels can allow the maintenance of ESC in a 3D scaffold that recapitulates their physiological milieu and therefore removes the need for exogenous animal-derived feeder cells.

Alginate gels are commonly used as a cell culture scaffold for ESC as this material supports the differentiation[47–57] of these cells and due to immune-isolatory properties,[58] eliminates formation of teratomas that are prone to develop with the transplantation of ESC. Alginate scaffolds were previously used to direct the differentiation of ESC into insulin-producing cells,[57] neural cells,[48] cardiac cells,[49] hepatocytes,[54] osteogenic and chondrocyte cells,[50-53] definitive endoderm,[47] and embryonic bodies.[55,56] The generation of these different cell types and tissues is dependent on the porous, mechanically strong and protective environment provided by alginate gels.

Porous alginate gel scaffolds were reported to mediate the formation of human embryoid bodies (hEB) from human ESC.[56] The physical constraints of the gel encouraged small, round hEB and induced vasculogenesis in the forming hEB to a greater extent than in static or rotating cultures. Furthermore, under biochemical conditions designed to direct the differentiation of ESC to insulin-producing cells,[57] hepatocytes,[54] and neural cells,[48] alginate encapsulation was shown to maintain and enhance the phenotype of differentiated cells. Interestingly, a report showed that ESC differentiation could be directed away from the hepatocyte lineage and towards the neural lineage by physical cell–cell aggregation blocking in alginate gels, and that 2.2% (w/v) alginate microencapsulation could be optimally adapted to ESC neural differentiation.[48]

The alginate culture system for ESC differentiation is amenable to tissue engineering with the use of *in vitro* bioreactors.[49,50] A recent study demonstrated that human ESC encapsulated in poly-L-lysine (pLL)-coated alginate capsules and pLL-layered liquid core (LC) alginate beads were directed along cardiomyogenic lineage in stirred-suspension bioreactors.[49] Another report described the generation of 3D mineralized constructs that displayed the morphological, phenotypic, and molecular attributes of the osteogenic lineage, as well as mechanical strength and mineralized calcium/phosphate deposition.[50] This was suggested as a bioprocess that could be automated, scaled up and applied to bone tissue engineering to produce macroscopic bone or cardiac tissue.

The potential use of ESC for tissue engineering and therapeutic delivery could be managed with the bioprocessing and storage of these cells in the same alginate scaffolds that support their differentiation and their delivery to damaged tissues. Alginate microcapsules were reported as an efficient methodology for the proliferation of ESC in stirred tank bioreactors, before the subsequent storage (by cryopreservation) of these cells immobilized in the gel.[59,60] Cryopreservation of stem cells, especially ESC, is problematic because of low post-thaw cell survival rates and spontaneous differentiation following recovery.[59] The viability of mouse ESC encapsulated in arginine–glycine–aspartic acid–serine (RGDS)-coupled calcium alginates (1.2% (w/v)), was described as significantly higher than those in suspension or in unmodified alginates, after cryopreservation. Another study demonstrated an approximately 20-fold increase in cell concentration with this culture method and high cell recovery yields after cryopreservation.[60] This second study showed that microencapsulation improved the culture of ESC aggregates by protecting cells from hydrodynamic shear stress, by controlling aggregate size and by maintaining cell pluripotency. Alginate gels may also be used to store ESC under ambient conditions in sealed containers.[61] Under these unusual storage conditions, ESC were 74% viable and the progenitor phenotype of these cells was preserved. Moreover, this novel method for storage/transport of these stem cells was comparable to conventional cryopreservation methodology, indicating the possibility for more economical and less logistically challenging storage approaches for ESC.

A key advantage of using alginate gels to culture ESC is that the conditions for directing differentiation of ESC in these gels do not require feeder systems.[62,63] In a recent study ESC encapsulated in 1.1% (w/v) calcium alginate hydrogels could be grown in basic maintenance medium, without a feeder layer, for a period of up to 260 days.[62] Cell aggregates formed within the hydrogels did not form any of the three germ layers, but ESC retained their pluripotency (gene markers characteristic of pluripotency Oct-4, Nanog, SSEA-4, TRA-1-60 and TRA-1-81 were expressed) and could differentiate when they were subsequently cultured in a conditioned environment. A porous chitosan–alginate composite gel scaffold was also demonstrated to maintain self-renewal of ESC without the support of feeder cells or conditioned medium.[63]

Taken together, these studies show that alginate is clearly a versatile tool for the culture of ESC; this encourages speculation into clinical delivery of these cells to humans.

7.2.3 Engineering Clinically Viable Trabecular Bone and Cartilage Using Alginate Gel Scaffolds

7.2.3.1 Mesenchymal Stem Cells

The need for bone repair or replacement due to skeletal diseases, congenital malformations, trauma, and tumour resections is mainly addressed through the use of biomaterial scaffolds.[64] Mesenchymal stem cells (MSC) are progenitor cells capable of self-renewal and multi-lineage differentiation, and are recognized as an important cell source for cartilage repair and bone tissue engineering.[65] Alginate hydrogels are established as the 'scaffold of choice' for the culture and therapeutic delivery of MSC, and are particularly appropriate as MSC are hypoimmunogenic.[66] These biomaterials are applied as substrates that support the differentiation of MSC to osteogenic phenotype. A number of key studies demonstrate that bone formation in alginate gel scaffolds can be directed[67,68] and MSC immobilized in alginate gels can also mediate the formation of blood vessels.[69] Therefore, sophisticated technology may be developed that involves alginate–MSC scaffolds with intrinsic mechanical and chemical cues which drive the formation of trabecular or compact bone that is both vascularized and constructed for purpose.

A number of studies have been performed to develop optimized culture conditions for MSC in alginate gels in order to deliver these cells in a manner that promotes tissue regeneration or formation. The position of MSC in alginate gels was demonstrated to affect the differentiation of human MSC under bioreactor conditions.[70] Mineralization was observed to occur to differential extents; 79 ± 29% and 53 ± 25% in the inner and outer annuli of alginate microbead scaffolds respectively. Mechanical loading of MSC in alginate gels is another factor investigated to develop culture conditions appropriate for bone tissue engineering. Calcium signalling is implicated as a mediator in mechanotransduction pathways, therefore levels of this molecule were measured in MSC embedded within 4% alginate hydrogel constructs.[71] The frequency of calcium transients in alginate-encapsulated MSC was up- and down-regulated, with the application of a 20% static uniaxial compressive strain for 20 min, delivered after a 20 min unstrained period. The shape of alginate gels was also found to determine their effects on encapsulated MSC.[72] Microparticles with a surrounding protective alginate–poly-L-lysine–alginate membrane supported the viability of MSC more robustly than alginate microspheres.

Chondrogenesis is dependent on the recruitment of MSC that differentiate into chondroblasts which synthesize cartilage extracellular matrix (ECM).[73] Therefore the development of culture conditions for the chondrogenic differentiation of MSC in alginate gels is a clear priority. This was previously achieved by altering the 3D structure of this gel as well as through the use of mechanical stresses. Low-intensity ultrasound was shown to inhibit apoptosis and promote chondrogenesis of MSC in alginate gels.[74] Alginate beads and pellets that are established systems for the differentiation of MSC into chondrocytes were shown to induce this process in a differential manner.[75] The typical gene marker for articular cartilage, collagen type II, was more strongly expressed in the alginate bead system than in alginate pellets. Moreover, a recent study demonstrated that a disc-shaped, self-gelling alginate hydrogel could be used as a scaffold for chondrogenic differentiation of human bone marrow-derived MSC.[76]

Further to studies exploring suitable culture conditions for MSC in alginate gels, various strategies have been employed for the engineering and repair of bone and cartilage using MSC–alginate scaffolds. These have involved the use of materials established for bone repair, *e.g.* calcium phosphate cement (a bioactive scaffold that sets *in situ* and forms bioresorbable hydroxyapatite that bonds to bone),[77,78] together with alginate composite gels and chemically-modified alginate gels. Moreover, methods for the delivery of MSC encapsulated in alginate gels include using injectable scaffolds which mould into the shape of a cavity, as well as the incorporation of alginate beads containing MSC into calcium phosphate cement have proven suitable for the repair and formation of bone and cartilage.

A frequent approach for the engineering of bone is the construction of composite biomaterial systems consisting of alginate. Porous scaffolds made of β-tricalcium phosphate (β-TCP) were used to provide support for the formation of bone or osteogenic tissue in alginate hydrogels cultured in dynamic *in vitro*[79] conditions and *in vivo*, implanted subcutaneously in mice.[80] A report demonstrated that a 3D culture system consisting of a hybrid sponge composite of β-tricalcium phosphate–alginate–gelatin induced osteogenic differentiation of rat bone marrow-derived MSC.[81] A less developed study described nucleation of a bone-like hydroxyapatite mineral that, when used to coat alginate, allowed the attachment and proliferation of human MSC seeded on the surface of this scaffold.[82]

The repair of full-thickness articular cartilage defects and both *in vitro* and *in vivo* cartilage engineering have been attempted using MSC immobilized in alginate hydrogels and composite alginate scaffolds. Injured articular cartilage has a limited intrinsic capacity for repair. The formation of neocartilage is mediated by MSC; injuries that penetrate the subchondral bone of normal articular cartilage trigger the migration of MSC from bone marrow to the injured area where they generate neocartilage. *In vivo*, this repair process is not efficient, however, due to the formation of fibrocartilage. Native alginate gels and hydrophobically-modified (covalent fixation of octadecyl chains onto the polysaccharide backbone by esterification) alginate gels were successful in the treatment of cartilage injuries.[83] Alginate gels were also reported to promote the formation of cartilage matrix proteins from encapsulated MSC, and MSC–alginate constructs that were transplanted beneath the dorsal skin of nude mice produced hyaline cartilage.[84] Another *in vivo* study showed that a polylactic acid–alginate amalgam seeded with bone marrow-derived MSC and stimulated *in vitro* with TGF-β filled osteochondral defects in the canine femoral condyle with a cartilaginous tissue cartilage-like matrix after a 6 week treatment period.[85]

MSC immobilized in alginate gels are not only applied to the regeneration and engineering of bone and cartilage. Studies have shown that MSC in alginate gels can differentiate to a hepatocyte phenotype,[86] indicating that these cells cultured in alginate scaffolds may be applicable to hepatic tissue engineering. Alginate-encapsulated MSC may also be viable therapy for damaged cardiac tissue following myocardial infarction (MI).[69,87] Human MSC in alginate microspheres were shown to reduce infarct area and enhance arteriole formation in animal models following MI events.[87] This may be due to the ability of MSC in alginate gels to secrete angiogenic factors and induce angiogenesis.[69]

The next step in the use of specialized MSC–alginate niche-like environments is to form ordered, stratified cartilage and bone.

7.2.3.2 Chondrocytes

The interaction between chondrocytes and their surrounding ECM plays an important role in regulating cartilage metabolism in response to environmental cues. The formation of new cartilage *in vitro* for tissue repair requires an environment that will provide these conditions. Appropriate conditions for chondrocyte culture and differentiation in alginate gels have been developed. The tendency of chondrocytes cultured in a monolayer to dedifferentiate has been addressed through the use of 3D matrices that maintain the phenotype of these cells.[88] Articular chondrocytes embedded in alginate gel produce, *de novo*, a matrix rich in collagens and proteoglycans.[89] When released from this gel, chondrocytes are surrounded by a tightly bound cell-associated matrix, which is similar to the pericellular and territorial matrices identified in cartilage.

In addition to remaining functional and viable within alginate gels, chondrocytes retain their normal metabolic functions. Microdialysis was used to detect local changes in the metabolism of bovine articular chondrocytes entrapped in an alginate gel and cultured in a bioreactor for 2 weeks.[90] Concentration gradients of glucose and lactose (major metabolites produced) within the construct were evident, with the highest lactate concentrations in the construct centre. Culture conditions have included continuous hydrostatic pressure on chondrocytes embedded in alginate gels as a way to induce or improve their ability to form cartilage.[91] Other approaches have described the generation of stratified cartilage in alginate gels by layering chondrocyte-seeded alginate scaffolds.[92]

Alginate gels as fillers for the repair of cartilage defects are well characterized. This filler methodology has been applied to the formation of bone *in vivo*.[93] Bioreactor spaces were created in the tibia of New Zealand White rabbits and a biocompatible calcium alginate gel that crosslinked *in situ* was injected into this space. After 12 weeks, the bioreactor space was reconstituted by functional living bone. The engineered bone was easily harvested without any apparent postoperative morbidity to the subject and was transplanted successfully in a contralateral defect.

The use of alginate–chondrocyte constructs to repair craniofacial injuries is particularly important because alginate is easily malleable and can therefore be used to mould the right shape of cartilage. Injectable tissue-engineered autologous cartilage is promising for potential use in oral and maxillofacial surgery. Studies exploring this have demonstrated the restoration of craniofacial cartilage using polyvinyl alcohol–alginate gels.[94] A study that may represent an important milestone in reconstructive surgery for severe facial injuries involved the development of a simple method to create complex structures with good 3D tolerance in order to form cartilage in specific shapes.[95] Chondrocytes were harvested from sheep auricular cartilage and suspended in 2% alginate. The mixture of cells and gel was injected into Silastic moulds and implanted subcutaneously in the necks of the sheep from which the cells had originally been harvested. Analyses of the implanted constructs indicated cartilage formation with 3D shape retention. A recent study described a rapid-curing alginate gel system constructed by inducing gelation of a 2% (w/v) solution of a high-G alginate (65–75% G) with $CaCl_2$.[96] The *in vitro* cell culture of chondrocytes in the gel resulted in alginate–cell constructs that lacked the continuous, interconnected collagen–proteoglycan network of hyaline cartilage, but the gel system was capable of supporting periosteum-derived chondrogenesis.

Investigations of the viscosity and stiffness of alginate gels is essential to determine whether or not these structures are suitable for the engineering of cartilage that needs to be elastic for load bearing. Dynamic loading and perfusion culture environments alone were shown to enhance cartilage ECM production in dedifferentiated articular chondrocytes.[97] These factors are beginning to be studied to determine if they could enhance the generation of cartilage ECM from adult human articular chondrocytes embedded in alginate gels.[97]

In addition to cartilage engineering, the production of trabecular bone replacements is performed using chondrocytes embedded in alginate gels. Chondrocytes together with osteoblasts in alginate gels may allow the production of osteochondral implants that do not suffer from poor tissue formation and compromised integration. Studies employing this approach have been encouraging.[98,99] Arginine–glycine–aspartic acid (RGD)-modified alginate hydrogels crosslinked with bioactive strontium, zinc ions and calcium were described in a previous study for the culture of Saos-2 osteoblast-like cells.[98] Strontium gels produced from high-G alginates were slow to degrade, whereas those made with alginate rich in mannuronic acid (high M) degraded more quickly, and supported proliferation of Saos-2 osteoblast-like cells. Furthermore, chondrocytes and osteoblasts co-transplanted on hydrogels modified with an RGD-containing peptide sequence formed new bone tissue that grew in mass and cellularity by endochondral ossification.[99] The tissue was formed in a manner similar to normal longbone growth, and was organized into structures that morphologically and functionally resembled growth plates.

Collectively, these studies indicate that similar to studies applying MSC–alginate constructs to repairing and engineering cartilage and bone, chondrocyte–alginate constructs are on the cusp of preclinical use.

7.2.4 Alginate Gels for Cardiac Tissue Repair: Development of the Cardiac Patch

The myocardium is unable to regenerate because cardiomyocytes cannot replicate after injury. The heart is therefore an attractive target for tissue engineering to replace infarcted myocardium and enhance cardiac function. Cardiac tissue engineering is an alternative approach for the repair of damaged myocardium that aims to regenerate damaged myocardial tissues by applying heart patches created *in vitro*. The alginate cardiac patch for the bioengineering of cardiac tissue has proven successful for the repair of damaged myocardium.

A number of studies have validated the efficacy of the cardiac patch for the repair of infarcted hearts. One study showed that fetal cardiac cells seeded within porous alginate scaffolds formed multicellular contracting aggregates within the scaffold pores, attenuated left ventricle (LV) dilatation, and induced no change in LV contractility in rats following MI.[100] The engineering of a cardiac patch with mature vasculature was also recently attempted.[101] The patch was constructed by seeding neonatal cardiac cells with a mixture of prosurvival and angiogenic factors into an alginate scaffold capable of factor binding and sustained release. After 48 h in culture, the patch was vascularized for 7 days on the omentum, then explanted and transplanted onto infarcted rat hearts, 7 days after MI induction. When evaluated 28 days later, the vascularized cardiac patch showed structural and electrical integration into host myocardium.

Alginate cardiac patches may be used for the delivery of therapeutic proteins applied to induce myocardial regeneration after MI.[102,103] The release of insulin-like growth factor 1 (IGF-1) followed by hepatocyte growth factor (HGF) from affinity-binding alginate was recently reported.[102] This was performed in a rat model of acute MI; an intramyocardial injection of the dual IGF-1/HGF affinity-bound alginate biomaterial preserved scar thickness, attenuated infarct expansion and reduced scar fibrosis after 4 weeks, together with increased angiogenesis and mature blood vessel formation at the infarct. A similar study described the design of alginate gels with affinity-binding moieties that enabled the binding of heparin-binding proteins and their controlled presentation and release.[103] The modified gel served as a reservoir for IGF-1 and HGF, and led to improvements in cardiac structure and function, by thickening cardiac tissue scars and preventing LV remodelling and dilatation.

Injectable *in situ* gelling alginate is another approach that is under development; this was demonstrated as effective for treating old and recent infarcts in animal models.[104] A calcium-crosslinked alginate solution with low viscosity was injected into infarcts and underwent phase transition into a hydrogel. The biomaterial was replaced by connective tissue within 6 weeks. Serial echocardiography studies before and 60 days after injection showed that injection of alginate biomaterial into 7 day-old infarcts increased scar thickness and attenuated LV systolic and diastolic dilatation and dysfunction. A report showed that neonatal rat cardiac cells seeded into RGD-immobilized alginate scaffolds promoted cell adherence to the matrix, prevented cell apoptosis and accelerated cardiac tissue regeneration.[105] Within 6 days, the cardiomyocytes reorganized their myofibrils and reconstructed myofibres composed of multiple cardiomyocytes in a typical myofibre bundle.

A recent, novel approach to cardiac tissue engineering includes the use of tissue printing technology (TP) that offers the defined delivery of scaffolding materials and living cells.[106] TP, human cardiac-derived cardiomyocyte progenitor cells and alginate gels were combined to obtain a construct with cardiogenic potential for *in vitro* use or *in vivo* application. This approach allowed the generation of an *in vitro* tissue with homogenous distribution of cells in the scaffold. Printed cells were able to migrate from the alginate matrix and colonize a matrigel layer, forming tubular-like structures. These studies suggest that the use of alginate gels as cardiac bandages or patches may bring about dramatic changes in therapy for heart disease.

7.2.5 Alginate/Endothelial Progenitor Cell Platforms for Therapeutic Angiogenesis and Neovascularization

Endothelial progenitor cells (EPC) circulate in blood and differentiate into endothelial cells which line blood vessels.[107] EPC mediate the formation of new blood vessels (angiogenesis),[108] and, the formation of blood vessels from existing blood vessels (vasculogenesis).[109] Variations in the levels of these progenitor cells during heart disease were recently discovered. Mobilization of CD34$^+$ cells and EPC was shown to occur in heart failure, with elevation and depression of these cells in the early and advanced phases, respectively.[110] EPC were therefore indicated as cells which could be applied to therapeutic strategies against degenerative heart conditions. Transplantation of EPC indicated that these cells significantly improved blood flow, recovery and capillary density in animal models of hindlimb ischaemia.[111,112] EPC transplantation in MI patients also induced neovascularization and results from clinical trials on autologous EPC transplantation were promising.[113–115] The positive effects of transplanted EPCs in heart disease clinical trials were, however, modest and required repeated infusions of these cells.

The limitations encountered with direct EPC infusions may be eliminated by their local delivery using biomaterial carrier systems that promote tissue formation with the material as a scaffold. EPC delivered to the ischaemic limbs of mice within alginate cryogels were recently demonstrated to restore their function.[116] The premise of this study was that the population of EPC in the alginate gel would exit over a period of time, be incorporated into the damaged tissue and mediate regeneration of a vascular network. The study demonstrated that EPC together with outgrowth endothelial cells (OEC) improved engraftment of transplanted cells in ischaemic murine hindlimb musculature, and increased blood vessel densities from 260 to 670 vessels per mm^2, compared with direct cell injection. Furthermore, material deployment dramatically improved the efficacy of these cells in salvaging ischaemic murine limbs, whereas bolus OEC delivery was ineffective in preventing toe necrosis and foot loss.

Alginate gels have also been used as tools for the capture of EPC from blood.[117] This application is particularly important as the isolation and purification of EPC from circulating blood is not trivial and often results in the destruction and loss of many of these rare cells. The study demonstrated that four-arm amine-terminated poly(ethylene glycol) molecules together with antibodies within alginate hydrogels can enhance the ability of the hydrogels to capture EPC from whole human blood. The hydrogel coatings applied uniformly onto pillar structures within microfluidic channels could be easily dissolved with a chelator that allowed for effective recovery of EPC following capture.

Although the use of alginate gels as tools for the delivery of EPC is still at the very early stages of development, these initial studies provide confidence that this biomaterial may allow routine medical use of these progenitor cells.

7.2.6 The Construction of Neural Prosthetics and Culture Systems Using Alginate Gels

Neural stem cells (NSC) are multipotent cells with the capacity to differentiate into neurons, oligodendrocytes and astrocytes, the three major cell types which constitute the nervous system.[118,119] These cells are therefore of interest as therapy for a range of neuroregenerative applications.[120–122] NSC are able to promote neuroprotection and axonal regeneration of damaged nervous tissues, through secretion of neurotrophic factors such as nerve growth factor (NGF), brain-derived neurotrophic factor (BDNF), glial-derived neurotrophic factor (GDNF), and factors that inhibit axonal growth (matrix metalloprotease-2).[123,124]

Alginate hydrogels have been demonstrated as a 3D scaffold for the culture of NSC due to their biocompatibility with the central nervous system.[119,125–127] NSC producing neurotrophic factors can be encapsulated in alginate and localized to sites of injury as a therapeutic approach.[128] This strategy would be appropriate as alginate gels are established as efficient drug delivery substrates due to mass transfer properties which allow controlled release of small molecules.[15,129] The internal porosity of alginate gels may be a key factor that could be exploited in the therapeutic delivery of NSC using this material as a support.

The development of alginate-based culture conditions for NSC has included investigations into the impact of the physical properties of these constructs and modifications to their structure on NSC differentiation and delivery. Reports show that soft alginate gels support neurite outgrowth and protect neurons against oxidative stress.[130] Previous studies also show that alginate gels can be modified to the direct the growth of axons.[120] Alginate-based highly anisotropic capillary hydrogels were used to longitudinally direct regrowth of transected axons, a process that is essential for target reinnervation and functional recovery after spinal cord injury. Gels were used to support an *in vitro* axon outgrowth assay and the axon–alginate construct was subsequently used to treat adult rat spinal cord lesions *in vivo*.

Other modifications to alginate gels applied to NSC culture include the incorporation of poly(lactide-*co*-glycolide) microspheres loaded with alginate lyase into alginate hydrogels.[131] This modification allowed alginate hydrogels to be enzymatically degraded in a controlled and tunable manner. The expansion rate of NSC cultured *in vitro* in the degradable alginate hydrogel system increased significantly compared to those cultured in non-degrading alginate hydrogels. These studies have informed the engineering of neural tissue using alginate gels.[132] Parylene devices that included an open well seeded with NSC encapsulated in an alginate hydrogel scaffold were recently applied successfully as a prosthetic device.[132]

Alginate gels may be useful for the local delivery of neurotrophic factor secreting NSC, the engineering of neural tissue, and as a culture substrate that controls the growth and function of NSC.

7.3 The Influence of Alginate Gel Biophysical and Biochemical Properties on Cell Phenotype

The construction of a bioartificial niche or a bioreactor system requires careful assembly of components which define the *in vivo* surroundings of cells. The appropriate environmental cues must be provided for cells. Alginate gels containing cell adhesion peptides, *e.g.* RGD, provide a 3D environment that promotes cell adhesion, survival, migration and organization.

These gels have also been subject to modifications in porosity and mechanical stiffness, which allow effective transport of solutes and mimic physiological biomechanical stresses respectively. Increases in alginate gel porosity can enhance the elimination rates of soluble factors from cells and varying the mechanical strength of gels can control cell phenotype.

Altering the physical and chemical properties of alginate gels can be used as a means to maintain a population of viable cells as well as control cell behaviour. Cells may be induced to enhance the production of therapeutic factors whilst repressing the production of other factors that are less important therapeutically.

7.3.1 The Effect of Alginate Gel Biophysical Properties on Encapsulated Cells

The biomechanical and structural properties of alginate gels profoundly influence the behaviour of cells immobilized of cultured on the surface of these scaffolds (Table 7.1). The mass transfer of nutrients and oxygen to, and metabolic waste from cells encapsulated in alginate gels are determined by the shape,[133] stability[129] and M : G content[15,134] of gels.

Previous studies demonstrate that simultaneous cell encapsulation and pore generation inside alginate hydrogels is possible and increases in pore size enhance cell viability and metabolic activity.[25,36] Gelatin beads of 150–300 μm diameter were used as sacrificial porogens for generating pores within cell-laden alginate gels.[25] An encapsulated hepatocarcinoma cell line formed larger spheroids and higher albumin secretion in porous alginate hydrogels compared to non-porous conditions. Hydroxyethyl cellulose (HEC) was also demonstrated as an efficient live cell porogen in calcium alginate gels loaded with limbal epithelial cells (LEC).[36] The HEC polymer increased pore size in gels, and this increased LEC viability; gel stiffness decreased with greater amounts of HEC and this change in modulus correlated with high levels of cell viability.

In addition to internal porosity, changes in the strength of the alginate gel network caused by changes in the number of alginate strands held together in the 'egg-box' model were shown to affect the behaviour of encapsulated cells.[135] Increasing the concentration of $CaCl_2$ used at the time of gelation increased the strength of the alginate gel network and impeded the growth characteristics of bTC3 cells encapsulated in a high-G alginate. bTC3 cells encapsulated in a high-M alginate were not, however, affected by changes in $CaCl_2$ concentration because of the low percentage of consecutive G residues. The development of cell-interactive alginate gels with tunable degradation rates and mechanical stiffness by a combination of partial oxidation and bimodal molecular weight distribution has also been described.[136] Higher gel mechanical properties (13–45 kPa) was shown to increase myoblast adhesion, proliferation, and differentiation in a 2D cell culture model.

Table 7.1 The influence of biophysical modifications to the alginate structure on cell phenotype.

Cell type	Biophysical modification to alginate gel	Cell response to changes in alginate gel structure	Reference
Neural stem cells	Alginate hydrogels with variations in elastic moduli	The neuronal marker β-tubulin III was expressed at the greatest level in NSC within the softest alginate hydrogels, with an elastic modulus similar to that of brain tissues	Banerjee et al. (2009)[130]
Mesenchymal stem cells	Pellet and bead-shaped alginate hydrogels	The typical gene for articular cartilage, collagen type II was more strongly expressed in the alginate bead than the alginate pellet system. The specific gene for hypertrophic cartilage, collagen type X, was more rapidly expressed in the pellet system than the alginate bead system. The alginate bead system is more practical for differentiating MSC into articular chondrocytes	Yang et al. (2004)[75]
	Porous gelatin–alginate scaffolds	The porous gelatin–alginate scaffold was biocompatible for neovascularization and was bioresorbed completely *in vivo*, depending upon the crosslink density. MSC were able to attach and proliferate on the scaffolds, and the self-renewal potential of MSC cultures was similar during *in vitro* culture and *in vivo* implantation	Yang et al. (2009)[145]

(*continued*)

Table 7.1 (continued)

Cell type	Biophysical modification to alginate gel	Cell response to changes in alginate gel structure	Reference
Chondrocytes	Rapid-curing alginate gel system, formulated by inducing gelation of a 2% (w/v) solution of a high-G alginate (65–75% G) with a 75 mM solution of CaCl$_2$	Culture of chondrocytes in the gel yielded alginate–cell constructs without continuous, interconnected collagen–proteoglycan network of hyaline cartilage. Periosteum cultured *in vitro* within the gel, significant quantities (>50%) of the total area of the periosteal explants was composed of cartilage that contained cartilage-specific proteoglycans	Stevens *et al.* (2004)[96]
Limbal epithelial cells	Calcium alginate gels modified with the live cell porogen, hydroxylethyl cellulose	The viability of encapsulated limbal epithelial cells was influenced by HEC-mediated porosity and stiffness of alginate gels. LEC were optimally viable in 1.2 : 1.2% and 1.2 : 2.4% alginate–HEC gels	Wright *et al.* (2012)[36]

These studies may serve as a precursor for those investigating the structure of alginate for tissue engineering purposes. Alginate is frequently studied as a scaffold for intervertebral disc (IVD) repair, as it closely mimics mechanical and cell-adhesive properties of the nucleus pulposus (NP) of the IVD.[137] The relationship between alginate concentration and scaffold stiffness was previously examined to find preparation conditions where the viscoelastic behaviour mimics that of the NP. Discs prepared from 2% alginate closely matched the stiffness and loss tangent of NP. The biosynthetic phenotype of native cells isolated from NP was preserved in alginate matrices up to 4 weeks of culturing. The modulus of alginate gels has also impacted upon the rate of proliferation of NSC;[130] an increase in the modulus of the hydrogels decreased NSC proliferation. Expression of the neuronal marker β-tubulin III was increased in NSC in the softest hydrogels, which had an elastic modulus comparable to that of brain tissues.

Alginate scaffolds designed to improved cell attachment may also be applied to engineering tissues. A study described the preparation of alginate–hydroxyapatite (HAP) composite scaffolds, which incorporated HAP to improve the mechanical and cell-attachment properties of the scaffolds.[138] These scaffolds had a well-interconnected porous structure with an average pore size of 150 μm in diameter and over 82% porosity. Rat osteosarcoma UMR106 cells, an osteoblastic cell line, seeded in the scaffolds displayed better cell attachment to the 75/25 and 50/50 alginate–HAP composite scaffolds than to the pure alginate scaffold. Similarly, a report demonstrated the design of structurally modified alginate capsules for the encapsulation and transplantation of pancreatic islets.[139] High-viscosity sodium alginate (SA-HV), cellulose sulfate (CS), poly(methylene-co-guanidine) hydrochloride (PMCG), calcium chloride, and sodium chloride were included in capsules. The capsule size, mechanical strength, membrane thickness, and permeability could be precisely adjusted and quantified by the various components, and it was biocompatible, non-cytotoxic and potentially immunoisolatory for pancreatic islet cells.

Considered together, these studies indicate that generic parameters for mechanical and physical modifications to alginate gels may be developed as the basis for more specific cell immobilization and culture systems.

7.3.2 The Effect of Biochemically Modified Alginate Gels on Cells

Alginate hydrogels can be oxidized, acetylated and sulfated to construct environments appropriate to their use as cell culture, cell delivery or drug release systems (Table 7.2). Sulfation and oxidation improve the adhesion of cell-attachment motif peptides including RGD, YIGSR and glycosaminoglycans to gels.

Matrix-attached RGD peptide has been applied successfully in the engineering of cardiac tissue within macroporous alginate scaffolds. Neonatal rat cardiac cells were seeded into RGD-immobilized or unmodified alginate scaffolds.[105] The immobilized RGD peptide promoted cell adherence to the

Table 7.2 The effects of biochemical properties of alginate gels on cell phenotype.[a]

Cell type	Biochemical modification to alginate gel	Cell response to changes in alginate gel structure	Reference
Embryonic stem cells	pLL-coated alginate capsules with solid and LCs	Human ESC in pLL-layered LC alginate beads differentiated towards heart cells in serum-containing media. Higher fractions of cells expressing cardiac markers were detected in ESC encapsulated in LC than in solid beads	Jing et al. (2010)[49]
	RGDS-coupled calcium alginates (1.2% w/v)	Mouse ESC cryopreserved in RGDS-alginate beads had a higher expression of stem cell markers compared with cells cryopreserved in suspension or cells cryopreserved in unmodified alginates. The post-thaw cell viability was significantly higher for cells encapsulated in RGDS-modified alginate	Sambu et al. (2011)[59]
Neural stem cells	High-G (68%) and high-M (54%) alginates, with or without a pLL coating layer	BDNF, GDNF, and NGF from the encapsulated cells were detected from NSC encapsulated cells in a high-G alginate without pLL	Purcell et al. (2009)[128]
	PLGA microspheres loaded with alginate lyase into alginate hydrogels	A significant increase in the expansion rate of NPC cultured in degrading alginate hydrogels vs. NPC cultured in non-degrading (standard) alginate hydrogels was observed	Ashton et al. (2007)[131]

Cardiac progenitor cells	Alginate gels modified with the adhesion peptide G4RGDY and HBP G4SPPRRARVTY	The cardiac tissue developed in the HBP/RGD-attached scaffolds revealed the best features of a functional muscle tissue. The expression levels of a-actinin, N-cadherin and Cx-43, and an increase in Cx-43 with time, further supported the formation a contractile muscle tissue in the HBP/RGD-attached scaffolds	Sapir et al. (2011)[35]
Endothelial progenitor cells	Four-arm amine-terminated PEG molecules with antibodies within alginate hydrogels	Enhanced the ability of the alginate hydrogels to capture EPC from whole human blood	Hatch et al. (2011)[117]
Mesenchymal stem cells	RGD-coupled alginate microspheres	MSC immobilized within RGD-alginate microspheres differentiated into the osteogenic lineage. Immobilized MSC enhanced the ability of neighboring endothelial cells to form tubelike structures	Bidarra et al. (2010)[69]
	Alginate partially crosslinked with a matrix MMP cleavable peptide (PVGLIG)	MSC displayed an elongated morphology within PVGLIG/RGD-alginate hydrogels and formed cellular networks, suggesting that cells were able to structurally reorganize the matrix, through enzymatic hydrolysis of PVGLIG residues	Fonseca et al. (2011)[143]

(continued)

Table 7.2 (continued)

Cell type	Biochemical modification to alginate gel	Cell response to changes in alginate gel structure	Reference
Chondrocytes	Photo-crosslinkable alginate macromers prepared by reacting sodium alginate and 2-aminoethyl methacrylate in the presence of 1-ethyl-3-(3-dimethylaminopropyl)-carbodiimide hydrochloride and N-hydroxysuccinimide	Methacrylated alginate macromer and photocrosslinked alginate hydrogels exhibited low cytotoxicity when cultured with primary bovine chondrocytes	Jeon et al. (2009)[144]

[a]Cx-43, connexin-43; ESC, embryonic stem cells; G, guluronic acid; HBP, heparin-binding peptide; LC, liquid core; M, mannuronic acid; MMP, matrix metalloproteinase; NPC, neural progenitor cells; PEG, poly(ethylene glycol); PLGA, poly(lactide-co-glycolide); pLL, poly-L-lysine; PVGLIG, proline–valine–glycine–leucine–isoleucine–glycine; RGDS, arginine–glycine–aspartic acid–serine.

matrix, prevented cell apoptosis and accelerated cardiac tissue regeneration. By contrast, a BioLineRx BL-1040 myocardial implant, an *in situ* gelling alginate scaffold modified with RGD and YIGSR, or a non-specific peptide (RGE), were not effective for treating post-MI trauma.[140]

Chemical modifications to alginate gels can be used to induce changes in cell phenotype. An alginate–sulfate scaffold was designed for the presentation and sustained release of TGF-β1, and its effects on the chondrogenesis of human mesenchymal stem cells (hMSC) were examined.[141] When attached to matrix *via* affinity interactions with alginate sulfate, TGF-β1 loading was significantly greater and the initial release of this growth factor from the scaffold was attenuated compared to its burst release (>90%) from scaffolds lacking alginate sulfate. This indicated that sulfate modification mediated sustained release of the growth factor. The sustained TGF-β1 release induced MSC chondrogenic differentiation; differentiated chondrocytes with deposited collagen type II were seen within three weeks of *in vitro* hMSC seeding.

Synthetic RGE- and RGD-containing peptides conjugated to sodium alginate and seeded with bovine bone marrow stem cells (BMSC) were demonstrated to differentiate to the chondrocyte lineage.[142] BMSC spread specifically on RGD-modified surfaces. After 7 days in 3D gel culture, the chondrogenic supplements (TGF-β1 and dexamethasone) significantly stimulated chondrocytic gene expression (collagen II, aggrecan, and Sox-9) and matrix accumulation (collagen II and sGAG) in RGE-modified gels, but this response was inhibited in the RGD-modified gels.

These studies indicate that chemical modifications to alginate gels must be tailored to the cell type of interest and that generic alginate-based artificial systems alone are not useful for cell culture or therapeutic delivery.

7.4 Perspectives

The immunoisolatory properties of alginate gels presents this biomaterial as an obvious choice for the transplantation of living cells and tissue fragments that either release therapeutic factors and/or are incorporated into healthy tissue. This biomaterial allows for the exchange of nutrients, oxygen and stimuli across cell membranes, whereas antibodies and immune cells from the host that are larger than the capsule pore sizes are excluded.

Early preclinical trials applying alginate gels for the transplantation of allogeneic parathyroid and pancreatic tissue resulted in normal levels of parathormone and glucose respectively, and grafts functioned well for several months without the need for immunosuppressive therapy.[58] More recent preclinical work (Table 7.3) has demonstrated that alginate is suitable as a material support for a number of tissues including bone, myocardium and vascular tissue grafts. Bone can be generated *in situ* with alginate supports and myocardium and vascular tissue can be engineered *in vitro* in alginate gels. Therefore alginate gels may be used with the body as a bioreactor or a physiological environment can be reconstructed within these gels to engineer tissue and/or maintain cells in a functional state.

Table 7.3 Preclinical studies of therapeutic cell delivery using alginate gels.[a]

Cell type	Preclinical application of cells encapsulated in alginate gels	Therapeutic outcome	Reference
Embryonic stem cells	Intraperitoneal transplantation of alginate-encapsulated insulin-producing cells from the embryo-derived MEPI-1 line, in diabetic mice	Within days after transplantation, hyperglycaemia was reversed followed by approximately 2.5 months of normo- to moderate hypoglycemia before relapsing	Shao et al. (2011)[146]
Mesenchymal stem cells	Investigation of the efficacy of chitosan–alginate gel/mesenchymal stem cells/bone morphogenetic protein-2 composites as injectable materials for new bone formation	After injecting CA-MSCs-BMP into mice, subcutaneous nodules formed by 12 weeks; these were hard and resisted compression	Park et al. (2005)[147]
	Investigation of the use of hMSC encapsulated in RGD-modified alginate microspheres to mediate myocardial repair	Alginate microbeads and hMSC encapsulated in microbeads maintained LV shape and prevented negative LV remodelling after a MI in rats. Cell survival was significantly increased in the encapsulated hMSC group compared with phosphate buffered saline control or cells alone. Microspheres, hMSCs, and hMSCs in microsphere groups reduced infarct area and enhanced arteriole formation	Yu et al. (2010)[87]

Alginate Hydrogels for the 3D Culture and Therapeutic Delivery of Cells 161

Chondrocytes	To engineer bone, a cavity was created between the surface of a long bone (tibia) and the membrane rich in pluripotent cells that covers it, the periosteum, in New Zealand White rabbits. Controlled manipulation of the periosteal space was achieved by using a hydraulic elevation procedure. The cavity was filled with a calcium alginate gel, 200 mm^3 in volume	Cells proliferated within the inner layer of the periosteum and by day 3 the bioreactor space was filled with spindle-shaped periosteal cells and capillaries. Woven bone was formed at 2 weeks, and this then transformed into compact bone tissue. Bone harvested after 6 weeks and transplanted it into contralateral tibial defects was completely integration after 6 weeks with no apparent morbidity at the donor site	Stevens et al. (2005)[93]
Chondrocytes	Allogeneic implants of chondrocytes in alginate gels were tested for their ability to reconstruct artificially damaged articular rabbit cartilage *in vivo*. The suspensions of chondrocytes in alginate were gelled by the addition of CaCl$_2$ solution directly into the bone defects producing a construct *in situ* that directly inserted and adhered to the subchondral bone and to the walls of intact cartilage	Almost complete repair of the bone defect with normal tissue structure was observed 4–6 months after implantation of the chondrocytes. Controls in which calcium alginate alone was implanted developed only a fibrous cartilage	Fragonas et al. (2000)[148]

(*continued*)

Table 7.3 (continued)

Cell type	Preclinical application of cells encapsulated in alginate gels	Therapeutic outcome	Reference
Endothelial progenitor cells	Investigation of the ability of VEGF121-presenting alginate scaffolds to enhance the therapeutic efficacy of transplanted OEC and EPC. This construct was used to reduce ischaemic limb injuries in SCID mice that were subjected to femoral artery and vein ligation	Transplantation of OEC on scaffolds presenting VEGF121 prevented autoamputation and reduced levels of necrosis. Transplanting a combination of EPC and OEC on alginate scaffolds resulted in normal perfusion levels by 4 weeks	Silva et al. (2008)[116]
Neural stem cells	To assess the capacity to promote directed axon regeneration, alginate-based highly ACH were introduced into an axon outgrowth assay in vitro and adult rat spinal cord lesions in vivo	Axonal profiles immunoreactive for neurofilament, tyrosine hydroxylase grew robustly into alginate-based capillary gels in a longitudinal orientation, and neurofilament positive axons were aligned within the capillaries. ACH caused highly oriented robust axonal regeneration following spinal cord injury in adult rats	Prang et al. (2006)[120]

Neural stem cells	Investigation of the efficacy of a novel, cell-seeded cortical neural prosthesis constructed from fabricated parylene devices that included an open well seeded with NSC encapsulated in an alginate hydrogel scaffold. The prosthetic device was implanted into 16 male Sprague Dawley rats (300–350 g); four untethered probes (two containing NSC-seeded alginate, one with alginate only, and one untreated probe) were used	Neuronal loss and glial encapsulation associated with cell-seeded devices were attenuated during the initial week of implantation and increased by 6 weeks after transplantation compared to control conditions. The graft cells may secrete neuroprotective and neurotrophic factors that enhance the healing response, and subsequent cell death and scaffold degradation that may account for a reversal of tissue repair	Purcell et al. (2009)[128]
Cardiac progenitor cells	Investigation of a strategy for the engineering of a cardiac patch with mature vasculature by heterotopic transplantation onto the omentum. The patch was constructed by seeding neonatal cardiac cells with a mixture of prosurvival and angiogenic factors into an alginate scaffold	After 48 h in culture, the cardiac patch was vascularized for 7 days on the omentum, explanted and transplanted onto infarcted rat hearts, 7 days after induction of MI. 28 days later, the vascularized cardiac patch showed structural and electrical integration into host myocardium, induced thicker scars, and prevented further dilatation of the chamber or ventricular dysfunction	Dvir et al. (2009)[101]

[a]ACH, anisotropic capillary hydrogels; CA–MSCs–BMP, chitosan–alginate gel/MSC/bone morphogenetic protein; EPC, endothelial progenitor cells; hMSC, human mesenchymal stem cells; LV, left ventricle; OEC, outgrowth epithelial cells.

The concept of an alginate hydrogel as a basis for a bioartificial niche system applicable as a medical device or a platform for the study of cell function is already realized. But these systems are either too generic or they are exclusive to only one particular cell type (*e.g.* NSC and cardiac cells) that is under development. Specialized artificial niche systems for individual cell types may be constructed by exploring a number of biochemical and structural permutations to identify the most appropriate biomaterial or composite biomaterial.

Alginate gels may be translated as powerful biomedical tools in the clinic. Alginate-based medical devices may be tailored to treat varying extents of a disease, *i.e.* grafts may be used for the replacement of damaged tissue or the repair of tissue mediated by therapeutic factors released in a controlled manner from cells encapsulated in alginate gels.

References

1. K. I. Draget, O. Smidsrød and G. Skjåk-Bræk, in *Biopolymers*, ed. S. De Baets, E. J. Vandamme and A. Steinbuchel, Wiley, Weinheim, 2002, p. 215.
2. M. Rinaudo, in *Comprehensive Glycoscience from Chemistry to Systems Biology*, ed. J. P. Kalmerling, Elsevier, London, 2007, p. 691.
3. W. Sabra, A. P. Zeng and W. D. Deckwer, *Appl. Microbiol. Biotechnol.*, 2001, **56**, 325.
4. P. H. Calumpong, P. A. Maypa and M. Magbanua, *Hydrobiologia*, 1999, **398**, 211.
5. D. Leal, B. Matsuhiro, M. Rossi and F. Caruso, *Carbohydr. Res.*, 2008, **343**, 308.
6. M. R. Torres, A. P. Sousa, E. A. Silva-Filho, D. F. Melo, J. P. Feitosa, R. C. de Paula and M. G. Lima, *Carbohydr. Res.*, 2007, **342**, 2067.
7. L.-E. Rioux, S. L. Turgeon and M. Beaulieu, *Carbohydr. Polym.*, 2007, **69**, 530.
8. T. A. Fenoradosoa, G. Ali, C. Delattre, C. Laroche, E. Petit, A. Wadouachi and P. Michaud, *J. Appl. Phycol.*, 2010, **22**, 131.
9. W. Sabra and W. D. Deckwer, in *Polysaccharides—Structural Diversity and Functional Versatility*, ed. S. Dumitriu, Marcel Dekker, New York, 2005, p. 515.
10. A. Haug, B. Larsen and O. Smidsrød, *Acta Chem. Scand.*, 1966, **20**, 183.
11. G. T. Grant, E. R. Morris, D. A. Rees, P. J. C. Smith and D. Thom, *FEBS Lett.*, 1973, **32**, 195.
12. I. Braccini and S. Perez, *Biomacromolecules*, 2001, **2**, 1089.
13. E. Bergstrom, D. M. Goodall and I. T. Norton, in *Gums and Stabilisers for the Food Industry*, ed. G. O. Phillips, P. A. Williams and D. J. Wedlock, IRL Press, Oxford, 1990, p. 501.
14. G. Skjak-Braek, H. Grasdalen and O. Smidsqbd, *Carbohydr. Polym.*, 1989, **10**, 31.
15. C. K. Kuo and P. X. Ma, *Biomaterials*, 2001, **22**, 511.

16. K. I. Draget, K. Steinsvag, E. Onsoyen and O. Smidsrod, *Carbohydr. Polym.,* 1998, **35**, 1.
17. Y. A. Morch, I. Donati, B. L. Strand and G. Skjak-Braek, *Biomacromolecules,* 2006, **7**, 1471.
18. B. Thu, P. Bruheim, T. Espevik, O. Smidsrod, P. Soon-Shiong and G. Skjak-Braek, *Biomaterials,* 1996, **17**, 1069.
19. J. L. Drury, R. G. Dennis and D. J. Mooney, *Biomaterials,* 2004, **25**, 3187.
20. H. J. Kong and D. J. Mooney, *Cell Transplant.,* 2003, **12**, 779.
21. M. Ahearne, Y. Yang, A. J. El Haj, K. Y. Then and K. K. Liu, *J. R. Soc., Interface,* 2005, **2**, 455.
22. B. F. Ju and K. K. Liu, *Mech. Mater.,* 2002, **34**, 485.
23. K. K. Liu and B. F. Ju, *J. Phys. Appl. Phys.,* 2001, **34**, L91.
24. J. Y. Sun, X. Zhao, W. R. Illeperuma, O. Chaudhuri, K. H. Oh, D. J. Mooney, J. J. Vlassak and Z. Suo, *Nature,* 2012, **489**, 133.
25. C. M. Hwang, S. Sant, M. Masaeli, N. N. Kachouie, B. Zamanian, S.-H. Lee and A. Khademhosseini, *Biofabrication,* 2010, **2**, 1.
26. S. J. Seo, Y. J. Choi, T. Akaike, A. Higuchi and C. S. Cho, *Tissue Eng., Part A,* 2006, **12**, 33.
27. U. Schuldt and D. Hunkeler, *J. Microencapsulation,* 2007, **24**, 1.
28. W. Li, Y. Hayashida, Y. T. Chen and S. C. Tseng, *Cell Res.,* 2007, **17**, 26.
29. M. Davanger and A. Evensen, *Nature,* 1971, **229**, 560.
30. K. M. Meek, S. Dennis and S. Khan, *Biophys. J.,* 2003, **85**, 2205.
31. J. V. Jester, T. Moller-Pederson, J. Huang, C. M. Sax, W. T. Kays, H. D. Cavangh, W. M. Petroll and J. Piatigorsky, *J. Cell Sci.,* 1999, **112**, 613.
32. N. C. Joyce, *Prog. Retinal Eye Res.,* 2003, **22**, 359.
33. W. Liu, M. Griffith and F. Li, *J. Mater. Sci.: Mater. Med.,* 2008, **19**, 3365.
34. E. Ozturk, M. A. Ergun, Z. Ozturk, A. B. Nurozler, K. Kececi, N. Ozdemir and E. B. Denkbas, *Int. J. Artif. Organs,* 2006, **29**, 228.
35. Y. Sapir, O. Kryukov and S. Cohen, *Biomaterials,* 2011, **32**, 1838.
36. B. Wright, R. A. Cave, J. P. Cook, V. V. Khutoryanskiy, S. Mi, B. Chen, M. Leyland and C. J. Connon, *Regener. Med.,* 2012, **7**, 296.
37. M. Faustini, M. Bucco, G. Galeati, M. Spinaci, S. Villani, T. Chlapanidas, I. Ghidoni, D. Vigo and M. L. Torre, *Reprod. Domest. Anim.,* 2010, **45**, 359.
38. C. Tamponnet, S. Boisseau, P. N. Lirsac, J. N. Barbotin, C. Poujeol, M. Lievremont and M. Simonneau, *Appl. Microbiol. Biotechnol.,* 1990, **33**, 442.
39. T. Sakurai, M. Kimura and M. Sato, *Mol. Hum. Reprod.,* 2005, **11**, 325.
40. B. N. Dontchos, C. H. Coyle, N. J. Izzo, D. M. Didiano, J. C. Karpie, A. Logar and C. R. Chu, *J. Orthop. Res.,* 2008, **26**, 643.
41. A. Biswas and R. Hutchins, *Stem Cells Dev.,* 2007, **16**, 213.
42. G. N. Stacey, F. Cobo, A. Nieto, P. Talavera, L. Healy and A. Concha, *J. Biotechnol.,* 2006, **125**, 583.
43. T. E. Ludwig, M. E. Levenstein, J. M. Jones, W. T. Berggren, E. R. Mitchen, J. L. Frane, L. J. Crandall, C. A. Daigh, K. R. Conard, M. S. Piekarczyk, R. A. Llanas and J. A. Thomson, *Nat. Biotechnol.,* 2006, **24**, 185.

44. L. M. Hoffman and M. K. Carpenter, *Nat. Biotechnol.*, 2005, **23**, 699.
45. S. M. Dellatore, A. S. Garcia and W. M. Miller, *Curr. Opin. Biotechnol.*, 2008, **19**, 534.
46. A. F. Godier, D. Marolt, S. Gerecht, U. Tajnsek, T. P. Martens and G. Vunjak-Novkovic, *Birth Defects Res., Part C*, 2008, **84**, 335.
47. M. Chayosumrit, B. Tuch and K. Sidhu, *Biomaterials*, 2010, **31**, 505.
48. L. Li, A. E. Davidovich, J. M. Schloss, U. Chippada, R. R. Schloss, N. A. Langrana and M. L. Yarmush, *Biomaterials*, 2011, **32**, 4489.
49. D. Jing, A. Parikh and E. S. Tzanakakis, *Cell Transplant.*, 2010, **19**, 1397.
50. Y. S. Hwang, J. Cho, F. Tay, J. Y. Heng, R. Ho, S. G. Kazarian, D. R. Williams, A. R. Boccaccini, J. M. Polak and A. Mantalaris, *Biomaterials*, 2009, **30**, 499.
51. H. Tanaka, C. L. Murphy, C. Murphy, M. Kimura, S. Kawai and J. M. Polak, *J. Cell. Biochem.*, 2004, **15**, 454.
52. W. L. Randle, J. M. Cha, Y. S. Hwang, K. L. Chan, S. G. Kazarian, J. M. Polak and A. Mantalaris, *Tissue Eng.*, 2007, **13**, 2957.
53. L. Buttery, R. Bielby, D. Howard and K. Shakesheff, *Methods Mol. Biol.*, 2011, **695**, 281.
54. S. Fang, Y. D. Qiu, L. Mao, X. L. Shi, D. C. Yu and Y. T. Ding, *Acta Pharmacol. Sin.*, 2007, **28**, 1924.
55. J. P. Magyar, M. Nemir, E. Ehler, N. Suter, J. C. Perriard and H. M. Eppenberger, *Ann. N. Y. Acad. Sci.*, 2001, **944**, 135.
56. S. Gerecht-Nir, S. Cohen, A. Ziskind and J. Itskovitz-Eldor, *Biotechnol. Bioeng.*, 2004, **88**, 313.
57. N. Wang, G. Adams, L. Buttery, F. H. Falcone and S. Stolnik, *J. Biotechnol.*, 2009, **144**, 304.
58. H. Zimmermann, D. Zimmermann, R. Reuss, P. J. Feilen, B. Manz, A. Katsen, M. Weber, F. R. Ihmig, F. Ehrhart, P. Gessner, M. Behringer, A. Steinbach, L. H. Wegner, V. L. Sukhorukov, J. A. Vasquez, S. Schneider, M. M. Weber, F. Volke, R. Wolf and U. Zimmermann, *J. Mater. Sci.: Mater. Med.*, 2005, **16**, 491.
59. S. Sambu, X. Xu, H. A. Schiffer, Z. F. Cui and H. Ye, *CryoLetters*, 2011, **32**, 389.
60. M. Serra, C. Correia, R. Malpique, C. Brito, J. Jensen, P. Bjorquist, M. J. Carrondo and P. M. Alves, *PLoS One*, 2011, **6**, e23212.
61. B. Chen, B. Wright, R. Sahoo and C. J. Connon, *Tissue Eng., Part C*, 2013, **19**, 568.
62. N. Siti-Ismail, A. E. Bishop, J. M. Polak and A. Mantalaris, *Biomaterials*, 2008, **29**, 3946.
63. Z. Li, M. Leung, R. Hopper, R. Ellenbogen and M. Zhang, *Biomaterials*, 2010, **31**, 404.
64. S. Bose, M. Roy and A. Bandyopadhyay, *Trends Biotechnol.*, 2012, **30**, 546.
65. P. S. Frenette, S. Pinho, D. Lucas and C. Scheiermann, *Annu. Rev. Immunol.*, 2013, **31**, 285.
66. S. A. Jacobs, V. D. Roobrouck, C. M. Verfaillie and S. W. Van Gool, *Immunol. Cell Biol.*, 2013, **91**, 32.

67. X. T. Mo, S. C. Guo, H. Q. Xie, L. Deng, W. Z. Ziang, X. Q. Li and Z. M. Yang, *Bone,* 2009, **45**, 42.
68. T. Fuji, T. Anada, Y. Honda, Y. Shiwaku, H. Koike, S. Kamakura, K. Sasaki and O. Suzuki, *Tissue Eng., Part A,* 2009, **15**, 3525.
69. S. J. Bidarra, C. C. Barrias, M. A. Barbosa, R. Soares and P. L. Granja, *Biomacromolecules,* 2010, **11**, 1956.
70. A. B. Yeatts, E. M. Geibel, F. F. Fears and J. P. Fisher, *Biotechnol. Bioeng.,* 2012, **109**, 2381.
71. J. J. Campbell, D. L. Bader and D. A. Lee, *J. Appl. Biomater. Biomech.,* 2008, **6**, 9.
72. E. Trouche, S. Girod Fullana, C. Mias, C. Ceccaldi, F. Tortosa, M. H. Seguelas, D. Calise, A. Parini, D. Cussac and B. Sallerin, *Cell Transplant.,* 2010, **19**, 1623.
73. C. Matta and R. Zakany, *Front. Biosci.,* 2013, **5**, 305.
74. H. J. Lee, B. H. Choi, B. H. Min and S. R. Park, *Tissue Eng.,* 2007, **13**, 1049.
75. I. H. Yang, S. H. Kim, Y. H. Kim, H. J. Sun, S. J. Kim and J. W. Lee, *Yonsei Med. J.,* 2004, **45**, 891.
76. S. R. Herlofsen, A. M. Kuchler, J. E. Melvik and J. E. Brinchmann, *Tissue Eng., Part A,* 2011, **17**, 1003.
77. R. Z. LeGeros, *Clin. Mater.,* 1993, **14**, 65.
78. M. Bohner and G. Baroud, *Biomaterials,* 2005, **26**, 1553.
79. C. Weinand, I. Pomerantseva, C. M. Neville, R. Gupta, E. Weinberg, I. Shapiro, H. Abukawa, M. J. Troulis and J. P. Vacanti, *Bone,* 2006, **38**, 555.
80. T. Matsuno, Y. Hashimoto, S. Adachi, K. Omata, Y. Yoshitaka, Y. Ozeki, Y. Umezu, Y. Tabata, M. Nakamura and T. Satoh, *Dent. Mater. J.,* 2008, **27**, 827.
81. M. B. Eslaminejad, H. Mirzadeh, Y. Mohamadi and A. Nickmahzar, *J. Tissue Eng. Regener. Med.,* 2007, **1**, 417.
82. D. Suarez-Gonzalez, K. Barnhart, E. Saito, Jr, R. Vanderby, S. J. Hollister and W. L. Murphy, *J. Biomed. Mater. Res., Part A,* 2010, **95**, 222.
83. M. K. Ghahramanpoor, S. A. Hassani Najafabadi, M. Abdouss, F. Bagheri and M. Baghaban Eslaminejad, *J. Mater. Sci.: Mater. Med.,* 2011, **22**, 2365.
84. H. L. Ma, T. H. Chen, L. Low-Tone Ho and S. C. Hung, *J. Biomed. Mater. Res., Part A,* 2005, **74**, 439.
85. J. S. Wayne, C. L. McDowell, K. J. Shields and R. S. Tuan, *Tissue Eng.,* 2005, **11**, 953.
86. N. Lin, J. Lin, L. Bo, P. Weidong, S. Chen and R. Xu, *Cell Proliferation,* 2010, **43**, 427.
87. J. Yu, K. T. Du, Q. Fang, Y. Gu, S. S. Mihardja, R. E. Sievers, J. C. Wu and R. J. Lee, *Biomaterials,* 2010, **31**, 7012.
88. A. Jonitz, K. Lochner, K. Peters, A. Salamon, J. Pasold, B. Mueller-Hilke, D. Hansmann and R. Bader, *Connect. Tissue Res.,* 2011, **52**, 503.
89. H. J. Hauselmann, R. J. Fernandes, S. S. Mok, T. M. Schmid, J. A. Block, M. B. Aydelotte, K. E. Kuettner and E. J. Thonar, *J. Cell Sci.,* 1994, **107**, 17.

90. O. A. Boubriak, J. P. Urban and Z. Cui, *J. R. Soc., Interface,* 2006, **3**, 637.
91. A. Fioravanti, D. Benetti, G. Coppola and G. Collodel, *Clin. Exp. Rheumatol.,* 2005, **23**, 847.
92. C. S. Lee, J. P. Gleghorn, N. Won Choi, M. Cabodi, A. D. Stroock and L. J. Bonassar, *Biomaterials,* 2007, **28**, 2987.
93. M. M. Stevens, R. P. Marini, D. Schaefer, J. Aronson, R. Langer and V. P. Shastri, *Proc. Natl. Acad. Sci. U. S. A.,* 2005, **102**, 11450.
94. D. A. Bichara, X. Zhao, N. S. Hwang, H. Bodugoz-Senturk, M. J. Yaremchuk, M. A. Randolph and O. K. Muratoglu, *J. Surg. Res.,* 2010, **163**, 331.
95. S. C. Chang, G. Tobias, A. K. Roy, C. A. Vacanti and L. J. Bonassar, *Plast. Reconstr. Surg.,* 2003, **112**, 793.
96. M. M. Stevens, H. F. Qanadilo, R. Langer and V. Prasad Shastri, *Biomaterials,* 2004, **25**, 887.
97. S. P. Grogan, S. Sovani, C. Pauli, J. Chen, A. Hartmann, Jr, C. W. Colwell, M. Lotz and D. D'Lima, *Tissue Eng., Part A,* 2012, **18**, 1784.
98. E. S. Place, L. Rojo, E. Gentleman, J. P. Sardinha and M. M. Stevens, *Tissue Eng., Part A,* 2011, **17**, 2713.
99. E. Alsberg, K. W. Anderson, A. Albeiruti, J. A. Rowley and D. J. Mooney, *Proc. Natl. Acad. Sci. U. S. A.,* 2002, **99**, 12025.
100. J. Leor, S. Aboulafia-Etzion, A. Dar, L. Shapiro, I. M. Barbash, A. Battler, Y. Granot and S. Cohen, *Circulation,* 2000, **102**, III56.
101. T. Dvir, A. Kedem, E. Ruvinov, O. Levy, I. Freeman, N. Landa, R. Holbova, M. S. Feinberg, S. Dror, Y. Etzion, J. Leor and S. Cohen, *Proc. Natl. Acad. Sci. U. S. A.,* 2009, **106**, 14990.
102. N. Landa, L. Miller, M. S. Feinberg, R. Holbova, M. Shachar, I. Freeman, S. Cohen and J. Leor, *Circulation,* 2008, **117**, 1388.
103. E. Ruvinov, J. Leor and S. Cohen, *Biomaterials,* 2011, **32**, 565.
104. E. Ruvinov, T. Harel-Adar and S. Cohen, *J. Cardiovasc. Transl. Res.,* 2011, **4**, 559.
105. M. Shachar, O. Tsur-Gang, T. Dvir, J. Leor and S. Cohen, *Acta Biomater.,* 2011, 7, 152.
106. R. Gaetani, P. A. Doevendans, C. H. Metz, J. Alblas, E. Messina, A. Giocomello and J. P. Sluijter, *Biomaterials,* 2012, **33**, 1782.
107. T. Asahara, T. Murohara, A. Sullivan, M. Silver, R. van der Zee, T. Li, B. B. Witzenbichler, G. Schatteman and J. M. Isner, *Science,* 1997, **275**, 964.
108. D. Gao, D. J. Nolan, A. S. Mellick, K. Bambino, K. McDonnell and V. Mittal, *Science,* 2008, **319**, 195.
109. S. Kaushal, G. E. Amiel, K. J. Guleserian, O. M. Shapira, T. Perry, F. W. Sutherland, E. Rabkin, A. M. Moran, F. J. Schoen, A. Atala, S. Soker, J. Bischoff and J. E. Mayer Jr, *Nat. Med.,* 2001, 7, 1035.
110. M. Valgimigli, G. M. Rigolin, A. Fucili, M. D. Porta, O. Soukhomovskaia, P. Malagutti, A. M. Bugli, L. Z. Bragotti, G. Francolini, E. Mauro, G. Castoldi and R. Ferrari, *Circulation,* 2004, **110**, 1209.
111. C. Urbich, C. Heeschen, A. Aicher, E. Dernbach, A. M. Zeiher and S. Dimmeler, *Circulation,* 2003, **108**, 2511.

112. C. Kalka, H. Masuda, T. Takahashi, W. M. Kalka-Moll, M. Silver, M. Kearney, T. Li, J. M. Isner and T. Asahara, *Proc. Natl. Acad. Sci. U. S. A.*, 2000, **97**, 3422.
113. A. Kawamoto, M. Katayama, N. Handa, M. Kinoshita, H. Takano, M. Horii, K. Sadamoto, A. Yokoyama, T. Yamanaka, R. Onodera, A. Kuroda, R. Baba, Y. Kaneko, T. Tsukie, Y. Kurimoto, Y. Okada, Y. Kihara, S. Morioka, M. Fukushima and T. Asahara, *Stem Cells*, 2009, **27**, 2857.
114. B. Assmus, J. Honold, V. Schachinger, M. B. Britten, U. Fischer-Rasokat, R. Lehmann, C. Teupe, K. Pistorius, H. Martin, N. D. Abolmaali, T. Tonn, S. Dimmeler and A. M. Zeiher, *N. Engl. J. Med.*, 2006, **355**, 1222.
115. K. Lunde, S. Solheim, S. Aakhus, H. Arnesen, M. Abdelnoor, T. Egeland, K. Endresen, A. Ilebekk, A. Mangschau, J. G. Fjeld, H. J. Smith, E. Taraldsrud, H. K. Grogaard, R. Bjornerheim, M. Brekke, C. Muller, E. Hopp, A. Ragnarsson, J. E. Brinchmann and K. Forfang, *N. Engl. J. Med.*, 2006, **355**, 1199.
116. E. A. Silva, E. S. Kim, H. J. Kong and D. J. Mooney, *Proc. Natl. Acad. Sci. U. S. A.*, 2008, **105**, 14347.
117. A. Hatch, G. Hansmann and S. K. Murthy, *Langmuir*, 2011, **27**, 4257.
118. K. Lai, B. Kaspar, F. H. Gage and D. V. Schaffer, *Nat. Neurosci.*, 2003, **6**, 21.
119. X. Li, T. Liu, K. Song, L. Yao, D. Ge, C. Bao, X. Ma and Z. Cui, *Biotechnol. Prog.*, 2006, **22**, 1683.
120. P. Prang, R. Muller, A. Eljaouhari, K. Heckmann, W. Kunz, T. Weber, C. Faber, M. Vroeman, U. Bogdahn and N. Weidner, *Biomaterials*, 2006, **27**, 3560.
121. S. Thuret, L. D. Moon and F. H. Gage, *Nat. Rev. Neurosci.*, 2006, **7**, 628.
122. K. Pfeifer, M. Vroemen, M. Caioni, L. Aigner, U. Bogdahn and N. Weidner, *Regener. Med.*, 2006, **1**, 255.
123. J. Ourednik, V. Ourednik, W. P. Lynch, M. Schachner and E. Y. Snyder, *Nat. Biotechnol.*, 2002, **20**, 1103.
124. J. Llado, C. Haenggeli, N. J. Maragakis, E. Y. Snyder and J. D. Rothstein, *Mol. Cell. Neurosci.*, 2004, **27**, 322.
125. P. Lu, L. L. Jones, E. Y. Snyder and M. H. Tuszynski, *Exp. Neurol.*, 2003, **81**, 115.
126. M. Matyash, F. Despang, R. Mandal, D. Fiore, M. Gelinsky and C. Ikonomidou, *Tissue Eng., Part A*, 2012, **18**, 55.
127. L. N. Novikova, L. N. Novikov and J. O. Kellerth, *Curr. Opin. Neurol.*, 2003, **16**, 711.
128. E. K. Purcell, A. Singh and D. R. Kipke, *Tissue Eng., Part C*, 2009, **15**, 541.
129. R. H. Li, D. H. Altreuter and F. T. Gentile, *Biotechnol. Bioeng.*, 1996, **50**, 365.
130. A. Banerjee, M. Arha, S. Choudhary, R. S. Ashton, S. R. Bhatia, D. V. Schaffer and R. S. Kane, *Biomaterials*, 2009, **30**, 4695.
131. R. S. Ashton, A. Banerjee, S. Punyani, D. V. Schaffer and R. S. Kane, *Biomaterials*, 2007, **28**, 5518.

132. E. K. Purcell, J. P. Seymour, S. Yandamuri and D. R. Kipke, *J. Neural. Eng.*, 2009, **67**, 026005.
133. S. Kintzios, I. Yiakoumetis, G. Moschopoulos, O. Mangana, K. Nomikou and A. Simonian, *Biosens. Bioelectron.*, 2007, **23**, 543.
134. S. Veriter, J. Mergen, R. M. Goebbels, N. Aouassar, C. Gregoire, B. Jordan, P. Leveque, B. Gallez, P. Gianello and D. Dufrane, *Tissue Eng., Part A,* 2010, **16**, 1503.
135. N. E. Simpson, C. L. Stabler, C. P. Simpson, A. Sambanis and I. Constantinidis, *Biomaterials,* 2004, **25**, 2603.
136. T. Boontheekul, E. E. Hill, H. J. Kong and D. J. Mooney, *Tissue Eng.*, 2006, **13**, 1431.
137. J. L. Bron, L. A. Vonk, T. H. Smit and G. H. Koenderink, *J. Mech. Behav. Biomed. Mater.*, 2011, **4**, 1196.
138. H. R. Lin and Y. J. Yeh, *J. Biomed. Mater. Res., Part B,* 2004, **71**, 52.
139. I. Lacik, M. Brissova, A. V. Anilkumar, A. C. Powers and T. Wang, *J. Biomed. Mater. Res.*, 1998, **39**, 52.
140. O. Tsur-Gang, E. Ruvinov, N. Landa, R. Holbova, M. S. Feinberg, J. Leor and S. Cohen, *Biomaterials,* 2009, **30**, 189.
141. T. Re'em, Y. Kaminer-Israeli, E. Ruvinov and S. Cohen, *Biomaterials,* 2012, **33**, 751.
142. J. T. Connelly, A. J. Garcia and M. E. Levenston, *Biomaterials,* 2007, **28**, 1071.
143. K. B. Fonseca, S. J. Bidarra, M. J. Oliveira, P. L. Granja and C. C. Barrias, *Acta Biomater.*, 2011, 7, 1674.
144. O. Jeon, K. H. Bouhadir, J. M. Mansour and E. Alsberg, *Biomaterials,* 2009, **30**, 2724.
145. C. Yang, H. Frei, F. M. Rossi and H. M. Burt, *J. Tissue Eng. Regener. Med.*, 2009, **3**, 601.
146. S. Shao, Y. Gao, B. Xie, F. Xie, S. K. Lim and G. Li, *J. Endocrinol.*, 2011, **208**, 245.
147. D. J. Park, B. H. Choi, S. J. Zhu, J. Y. Huh, B. Y. Kim and S. H. Lee, *J. Craniomaxillofac. Surg.*, 2005, **33**, 50.
148. E. Fragonas, M. Valente, M. Pozzi-Mucelli, R. Toffanin, R. Rizzo, F. Silvestri and F. Vittur, *Biomaterials,* 2000, **21**, 795.

CHAPTER 8

Mechanical Characterization of Hydrogels and its Implications for Cellular Activities

SAMANTHA L. WILSON[a,b], MARK AHEARNE[c], ALICIA J. EL HAJ[a], AND YING YANG*[a]

[a] Institute of Science and Technology in Medicine, School of Medicine, Keele University, Stoke-on-Trent, ST4 7QB, UK; [b] Ophthalmology, Division of Clinical Neuroscience, University of Nottingham, Queen's Medical Centre Campus, NG7 2UH, UK; [c] Trinity Centre for Bioengineering, Trinity Biomedical Sciences Institute, Trinity College Dublin, Ireland
*E-mail: y.yang@keele.ac.uk

8.1 Introduction

One of the challenges faced by scientists and clinicians is to fabricate physiologically relevant three-dimensional (3D) culture models with controllable biochemical and biophysical properties that can provide an *in vitro* platform to develop and test new clinical therapies.[1] The use of hydrogels is among some of the more promising approaches for the development of culture models for use in tissue engineering and regenerative medicine. These biomaterials consist of a water-swollen network of crosslinked hydrophilic polymer chains. The limited availability of native tissues for transplantation and *in vitro* testing has propelled the need to develop new hydrogels that replicate native tissue extracellular matrix (ECM). Hydrogel materials can be fabricated from natural protein polymers such as collagen, fibrin, agarose, gelatine or alginate, or from synthetic polymers such as poly(ethylene glycol) (PEG), poly(vinyl acetate) (PVA) or poly(acrylic acid) (PAA). The choice of polymer is vital when determining the suitability of a particular hydrogel material for a given application.[2] Natural protein hydrogels are advantageous in that they provide native biochemical cues and are able to simulate many aspects of the natural ECM. Synthetic hydrogels are valuable in that more well-defined, easily tuneable structures and mechanical properties can be achieved in comparison to protein-based hydrogels. Frequently a combination of natural and synthetic hydrogels is utilized in order to more closely mimic the dynamic native culture environments that change in response to cellular behaviour. Hydrogels provide a popular method of culturing cells in a 3D environment as they provide a structure in which a tissue can develop. Hydrogels act as a temporary matrix that allows cells to grow, move and communicate. Their viscoelastic characteristics, biocompatibility, availability and their ability to be remodelled by cells make them a suitable material for tissue regeneration. Cell-seeded hydrogel constructs can also replicate the close contact/adhesion that occurs between cells and ECM. Hence the mechanical properties of the resulting hydrogel construct become a unique property that mutually affects the constructed hydrogel and the cells. Characterization of the mechanical properties of hydrogel constructs may ultimately have implications on cellular actives.

In vivo and *in vitro* the extracellular environment is vital in controlling cell health and provides both chemical and mechanical stimuli that influence cellular behaviour.[1,3] *In vivo* cells are organized into tissues and organs with complex mechanical and structural architectures.[3] Both endogenous and exogenous forces act upon the cells and their surrounding environment. Endogenous factors include cell–integrin binding to the ECM and cellular responses to soluble factors such as growth factors and cytokines. Exogenous forces include gravity, substrate stiffness, polarity and surface to volume ratios[1] and tissue-specific interactions including traction forces generated by cells.[4]

Many cell types can be described as 'anchorage dependent' in that, to remain viable, they require a substrate to attach to.[5] Most soft tissues including vascular, cardiac, dermal, muscle, brain, tendon and cornea consist of ECM combined with adherent cells that possess elastic or viscoelastic characteristics[5] *in vivo*. Tissue culture plastic and glass coverslips provide a relatively rigid microenvironment (lacking many mechanical and biophysical cues) for cells to be cultured *in vitro* when compared to cells in native tissues. *In vivo* it is the combination of the cellular microenvironment and chemical and physical cues that mediate cellular behaviour. These niche environments are often very difficult to replicate *in vitro*. Cellular behaviour can vary markedly based upon the mechanical properties of the culture substrate.[6] Cellular migration, adhesion, proliferation, migration and differentiation can all affect and be affected by the mechanical properties of a tissue.[3]

8.2 Hydrogel Characterization Techniques

The mechanical and viscoelastic properties of hydrogel materials are important parameters when considering their suitability for use with cells and tissues. These same properties are highly reliant on environmental factors such as temperature and pH, in addition to the cell activity. Thus it is imperative to be able to determine the material properties of hydrogel constructs under conditions similar to the *in situ* environment in which they will be utilized.[2] Many techniques exist to measure the mechanical properties of hydrogels, and are centred on theories of rubber elasticity and viscoelasticity.[2] In general, hydrogels can be considered to behave in a viscoelastic manner, meaning they exhibit both viscous and elastic characteristics. The relationship between stress and strain in viscoelastic materials is dependent on time. Frequently used methods for measuring the mechanical properties of hydrogels include tensile, compression or indentation techniques.

Tensile testing or strip extensiometry testing are the most frequently used methods for the mechanical characterization of hydrogels.[7] This test involves clamping the hydrogel between two grips and then stretching it (Figure 8.1A). The amount of force required to stretch the hydrogel is measured and plotted against the distance the hydrogel has been stretched. The force and distance can be used to determine the stress and strain applied to the hydrogels and from this the Young's modulus can be calculated. Other parameters such as the ultimate tensile strength and yield strength of the hydrogels may also be determined using this test although these would require testing the sample to failure. This test can also be used to determine the viscoelastic characteristics of the hydrogel by stretching the hydrogel a predetermined distance

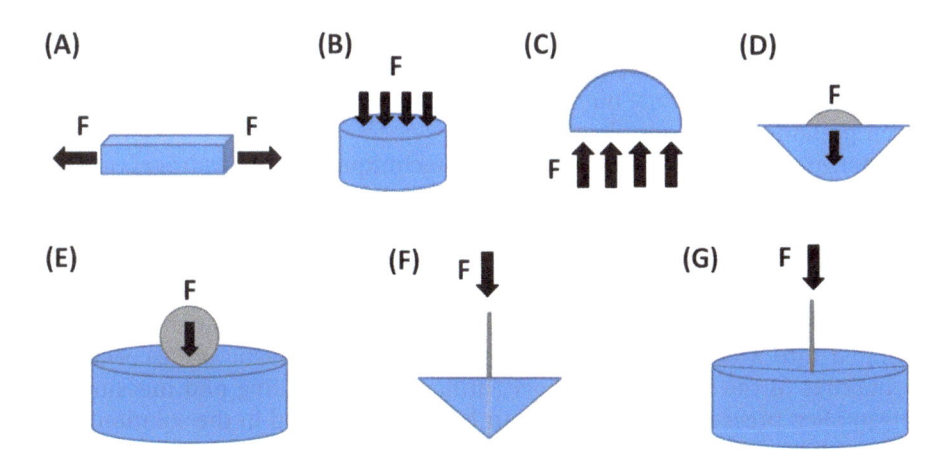

Figure 8.1 Schematic representation of the following mechanical tests where F represents the applied force: (A) tensile; (B) compression; (C) inflation; (D) spherical indentation (suspended); (E) spherical indentation (on substrate); (F) micro-indentation (suspended); (G) micro-indentation (on substrate).

and measuring the change in force required to maintain that elongation over time. If the hydrogel is viscoelastic, it should undergo stress relaxation resulting in a reduction in force over time until reaching equilibrium. The dynamic modulus of the hydrogels may also be determined by repetitively loading and unloading of the hydrogel. A variation on the standard tensile test involves using a hydrogel ring that is stretched between two posts. The advantage of this approach is that no grips are required. For both approaches hydrogels may be immersed in solution to ensure that they are maintained in a swollen state. The principle advantage of using these tests are their relative simplicity compared to other techniques. One limitation is that in general only uniform strips or rings can be tested; more complicated geometries would require more complex mechanical models to calculate values. In addition, the fragility of hydrogels can make them difficult to handle and grip in this system.

Compression testing has also been used to examine the mechanical properties of hydrogels. For this technique the hydrogel is placed under a uniform load that results in the hydrogel being compressed (Figure 8.1B). Depending on how the system is set up, either the load or the distance can be controlled while the other is measured. The resulting stress–strain relationship can be used to calculate the compressive modulus of the hydrogel. Due to the viscoelastic nature of hydrogels, typically dynamic moduli at specified frequencies or equilibrium modulus are determined. The equilibrium modulus is calculated from the stress–strain data after the hydrogel has undergone stress relaxation. This technique has previously been used to measure the mechanical properties of several types of cell-seeded hydrogels including fibrin, agarose and gellan gum hydrogels.[8] The samples can be fully submerged in solution during testing to prevent dehydration. Unlike extensometry, the geometry of the hydrogel is not limited to strips or rings, although a flat surface is required. Usually cylindrical hydrogel constructs are used. Limitations include bulging of the material and difficulties in applying even pressure to the sample.

A bulge or inflation test is a more novel technique where a hydrogel can be characterized by inflating it. The hydrogel is held in a sample holder and fluid is pumped underneath it causing it to bulge (Figure 8.1C). The bulge displacement as a function of the applied fluid pressure is measured using a charge-coupled device (CCD) camera or a laser.[7] The relationship between the applied pressure and the resultant strain on the hydrogel can be incorporated into a mathematical model to calculate the elastic or viscoelastic properties of the hydrogel. Leakage, difficulties controlling and measuring the applied pressure and dissolved air becoming trapped in the solution are all problems associated with bulge and inflation testing. The test is also only suitable for flat uniform hydrogels.

Indentation techniques have been widely used to characterize soft biomaterials including hydrogels. Hayes *et al.*[9] were one of the first groups to use indentation to examine the mechanical properties of a tissue. They used indentation to examine the mechanical properties of human cartilage.

Indentation has also been used to examine the adhesive characteristics of tissues[10] by measuring the adhesion force between the indenter and the tissue. There are several variations of the indentation techniques used to characterize hydrogels including spherical indentation, micro-indentation and nano-indentation.

Spherical indentation involves suspending a thin circular hydrogel around its outer circumferences in a specifically designed sample holder and placing a ball of known weight and size onto the hydrogel (Figure 8.1D). The weight of the ball causes the hydrogel to deform. The deformation is recorded *via* a CCD camera and the depth of indentation is used to calculate the elastic modulus of the hydrogel.[11] The viscoelastic properties of the hydrogel can also be monitored by measuring the change in central deformation over time.[12] This approach is particularly suitable for cell-seeded hydrogels as the whole assembly can be fully submerged in solution and be kept in an incubator at 37 °C while testing. This technique has been used to examine the effect of fibroblasts on the mechanical properties of collagen hydrogels[13] and the influence of nanofibres on hydrogel properties,[14] and for optimizing crosslinking conditions for hydrogels.[15] This technique allows for online, real-time and non-destructive measurements to be taken over prolonged culture periods.

A variation of the spherical indentation technique involves placing a hydrogel onto a flat substrate rather than suspending it (Figure 8.1E). This approach is more suitable to thicker hydrogels while the suspension approach is more suited to thinner hydrogels. The weight of the ball causes the hydrogel to deform and the deformation depth and weight of the ball can be applied to mathematical model called the Hertz model, to calculate the elastic modulus of the hydrogel. The main difficulty with this technique is accurately measuring the depth of indentation. One method around this problem is to use optical coherence tomography (OCT). The combination of spherical indentation and OCT has previously been used to measure the mechanical properties of agarose hydrogels.[16]

Micro-indentation involves deforming a hydrogel using a rigid indenter connected to a force transducer. The indenter is lowered onto the hydrogel and deforms it to a particular depth. The depth of indentation and the amount of force applied are both applied to theoretical model to calculate the mechanical properties of the hydrogel. The hydrogel may be suspend around it outer edge or placed flat on a substrate (Figure 8.1F and G). For suspended hydrogels a number of different theoretical models can be used to calculate the mechanical properties of the hydrogel.[17] For hydrogels on a flat substrate, the Hertz model is used to calculate the modulus of the hydrogel.[18] Micro-indentation can be used to examine regional variation across different areas of a hydrogel.

Nano-indentation works on the same principle as micro-indentation, but the tip size and indentation depth are on a nanometric scale. This apparatus consists of a sharp-tipped indenter attached to a cantilever beam. Mechanical characterization at this scale is limited to producing data on the surface properties of the hydrogel. The difficulties associated with using nanoindentation with hydrogels include accurate calibration of the instrument, applying a suitable mechanical model and the elimination of other sources of error.[19]

Eastwood et al.[20] developed a tensioning-culture force monitor system which can apply a predetermined loading to the hydrogel, in particular collagen hydrogel, and monitor the contraction strain generated by resident fibroblasts. The beauty of this system is that it can monitor the strain development for days continuously and visualize the associated global morphology change.

Ultrasound elastography is a technique that works by transmitting ultrasonic waves through the hydrogel and then reading backscattered waves, which can be used to form 2D images. When a force is applied to the hydrogel, the resulting displacement can be detected throughout the hydrogel. This information can then be processed to characterize the mechanical properties of the hydrogel. Fromageau et al.[21] used several variations of this technique to measure the Young's modulus of PVA hydrogels. They found that elastography produced similar mechanical values to standard mechanical testing techniques. The main limitation with this technique is the costs involved in the purchasing and running of the ultrasound equipment.

Recently, Li et al.[22] explored a novel approach that utilizes a low-coherence interferometer to detect the laser-induced surface acoustic waves (SAW) from agar hydrogels to mimic soft tissues. This technique allows for rapid characterization of the elastic properties of soft biological tissues, and has the advantage of being a non-destructive technique.

There is a widespread demand for the development of non-destructive techniques that permit the continuous measurement of hydrogel constructs for prolonged culture periods. The use of non-destructive mechanical characterization techniques is extremely valuable in that they allow for changes in mechanical properties over time to be characterized. Such changes can then be more accurately linked to cell activity and remodelling of the hydrogel matrix. Among other techniques, micro-indentation, ultrasound elastography and the combination of OCT and surface acoustic wave or with indentation techniques are extremely powerful tools for the characterization of the mechanical properties of hydrogels or soft tissues.

8.3 Effect of Hydrogel Mechanical Properties on Cellular Activities

In most native tissues, anchoring cells attach to the surrounding ECM. This ECM provides an inner physical support and its composition, topography and stiffness provides biochemical and biophysical cues that are necessary to the development and maintenance of these tissues. Until recently, chemical regulators within the extracellular environment have primarily been investigated, with little emphasis regarding the influence of mechanical regulation.[3] Similar to surface chemistry, the mechanical properties affect the local behaviours of tissues and cells. Normally cells embedded in tissues are able to 'probe', 'feel' or 'sense' the elasticity or stiffness of their surrounding matrix[5,6] or substrate as they anchor and pull themselves along during cell migration. 'Stiffness' refers to the measure of a material's ability to resist deformation and this can change during physiological processes including embryonic development, wound healing and pathological conditions.[4] In the body, the magnitude of stiffness is vast, ranging from a few kPa in adipose tissue[23] to GPa in bone.[4,24] A wide variation in matrix stiffness along with biochemical signals influence focal-adhesion structures and the cytoskeleton.[5] Previous studies using cells committed to a particular lineage, especially fibroblasts, on floating collagen gels and wrinkling-silicone sheets also suggest some responsiveness to the physical state of the matrix.[25]

In addition to applying force to its surroundings, the cells themselves respond to the resistance of the surrounding environment.[26] As the physical conditions of tissues can be altered during pathological conditions, this can affect cellular behaviour and differentiation. Cells adapt their adhesions, cytoskeletal configuration and general morphology in response to changes in substrate resistance or stiffness.[5] For example, cells attached to stiff, rigid constructs will form stable focal adhesions, whereas cells attached to less stiff materials will have diffuse and dynamic adhesion complexes.[5,26] This can have a direct impact on cellular migration and proliferation, such as increased proliferation in cells seeded onto stiffer substrates.[6] This can be linked to cellular wound healing responses as, often, the granulation and change in mechanical properties of scar tissue[6] is related to cellular infiltration and remodelling. In general, cells appear to adhere, spread and survive better on stiffer materials, although there are exceptions to this including neutrophils, which do not appear to be affected by substrate stiffness,[27] and neurons, which actually show improved survival on softer materials.[28] Studies on fibroblasts cultured on hydrogels have demonstrated that substrate stiffness significantly alters ECM assembly, cell spreading and motility.[6,29]

The mechanical properties of hydrogels can have a profound influence on regulating the phenotypic behaviour of cells. This is most noticeable with stem cells, where variations in stiffness can promote differentiation towards different lineages. Engler *et al.*[30] showed that the ability of stem cells to differentiate towards specific lineages was dependent on the substrate stiffness of the materials on which the cells were cultured. They noted that neurogenic differentiation was optimal at a stiffness of 0.1–1 kPa, myogenic differentiation at 8–17 kPa and osteogenic differentiation at 25–40 kPa. In hydrogels the effect of the concentration, which is directly linked to mechanical properties, on the differentiation of neuronal stem has been investigated.[31] Phenotypic neuronal markers were up regulated when the hydrogel stiffness matched that of brain tissue. Bian *et al.*[32] found that the chondrogenic capacity of mesenchymal stem cells could be optimized by varying the stiffness of hyaluronic acid hydrogels. Likewise, Steward *et al.*[33]

found that differentiation of mesenchymal stem cells could be partially regulated by hydrogel stiffness towards osteogenic or chondrogenic lineages.

Cell adhesion can also play a significant role on the mechanical properties of hydrogels and is vital for many other processes such as matrix contraction and cell migration. Hydrogel stiffness is dependent on the polymer concentration and the crosslinking density. A consequence of increasing the polymer concentration often results in a subsequent increase in adhesion sites. Cell adhesion can have a profound effect on cell behaviour and cell phenotype. Trappmann *et al.*[34] showed that differing protein anchorage densities can be used to regulate stem cell fate. This finding is important as it demonstrates the interdependent relationship between matrix stiffness, binding site availability and cell phenotype. Of course not all hydrogels facilitate binding by cells. Steward *et al.*[35] compared the phenotypic and cytoskeletal behaviour of cells in a hydrogel that facilitates binding, *i.e.* fibrin, and cells in a hydrogel that lack binding sites, *i.e.* agarose. Cells in hydrogels that allowed binding had a spread morphology while those in hydrogels that lacked binding sites maintained a spherical morphology.

8.4 Effect of Cellular Activity on Hydrogel Properties

In addition to hydrogels having an effect on the behaviour of cells, reciprocally the cell activity within the hydrogels can affect their bulk mechanical properties. Similar phenomena are found in diseases such as those that affect connective tissues and alter the mechanical properties of the tissue. The result is often the formation of hard tumours or the generation of ulcers. In these cases, it is the cellular activities that regulate the ECM. Hence, cells seeded in hydrogels can affect their mechanical properties through several metabolic activities including digestion *via* the production of enzymes; proliferation; matrix synthesis; contraction; ECM deposition and crosslinking.

Enzymes are produced by cells to initiate matrix remodelling and cell migration. Among the most prominent enzymes produced by cells are a family of enzymes called matrix metalloproteinases (MMPs). Mauch *et al.*[36] showed that fibroblasts seeded in collagen hydrogels release MMPs, resulting in the degradation of the hydrogel matrix. The enzymatic degradation of hydrogels by cells can be easily controlled by the addition of MMP inhibitors and other reagents that can influence the cell–matrix mechanical relationship. In addition to preventing matrix degradation, the inhibition of MMPs may also prevent the contraction and remodelling of hydrogels. Ahearne *et al.*[13] found that exogenous addition of actin filament disruptive agent cyochalasin, to fibroblast-seeded collagen hydrogels reduced the fibroblasts' contractibility and capacity of adhesion to hydrogel, which significantly decreased the elastic modulus of the hydrogel construct (Figure 8.2).

One of the most prominent mechanisms by which cells affect the mechanical properties of hydrogels is through the production of ECM proteins. For example, hydrogels seeded with chondrocytes, mesenchymal

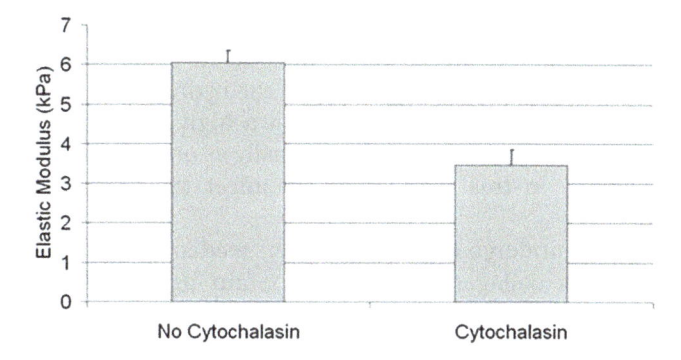

Figure 8.2 The elastic modulus change in response to the addition of cytochalasin in collagen hydrogel seeded with fibroblasts, which disrupted actin filament in fibroblasts.[13] Reproduced with permission from Mary Ann Libert Inc., previously printed in *Tissue Engineering Part C*.

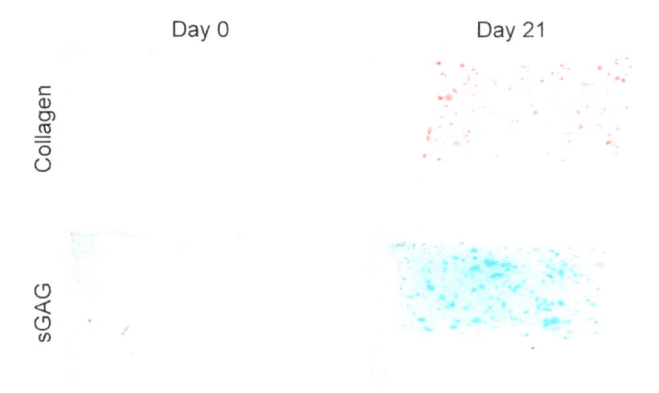

Figure 8.3 Increase in collagen (stained with picrosirius red) and sGAG (stained with Alcian blue) in cell-seeded agarose hydrogels after 21 days in culture in a chondrogenic medium.

stem cells or infrapatellar fat-pad derived progenitor cells have been shown to increase the stiffness of these hydrogels when they were cultured in a chemically defined prochondrogenic medium.[8,37–39] This increase in stiffness has been attributed to the release of collagen and sulfated glycosaminoglycans (sGAG) by cells that accumulate in the hydrogel and can be determined using biochemical analysis and histological staining, as shown in Figure 8.3. Hydrogels may also undergo calcification when cells are cultured in a pro-osteogenic medium. The increase in calcium deposition can increase the bulk stiffness of the hydrogel. Calcium phosphate may also be incorporated into the hydrogel prior to fabrication to increase the hydrogel stiffness or induce it to be formed. Douglas *et al.*[40] incorporated alkaline phosphatase into collagen and PEG-based hydrogels to induce their mineralization into calcium phosphate. They found that calcium phosphate was formed and mineralization increased the Young's modulus of the hydrogel. One factor that needs to be considered when relating hydrogel stiffness to ECM production is the ability of the hydrogel to retain the newly formed matrix components. Hydrogels with a high porosity or high water content may not retain matrix proteins as easily as other hydrogels. The loss of matrix proteins in this manner would affect the change in hydrogel stiffness.

Many hydrogels undergo contraction when seeded with cells (Figure 8.4). Hydrogels such as collagen and fibrin contain ligands that enable cell adhesion. Due to their limited mechanical strength, the cells are able to contract these hydrogels. Contraction can be a problem when attempting to design hydrogels that incorporate specific geometries that replicate the native tissues. These geometries can be destroyed by cells contracting their surrounding matrix.[41] Bell et al.[42] have suggested that contraction can be advantageous and may be used to enable the cells to engineer hydrogels into tissue-like structures. It has previously been shown that increasing the stiffness of these hydrogels reduces the rate of contraction. Ahearne et al.[43] showed that following 25 days in culture, collagen hydrogels seeded with fibroblasts at a concentration of 2.5 mg mL^{-1} reduced in thickness by 85% as a result of contraction, while stiffer hydrogels fabricated at a concentration of 4.5 mg mL^{-1} showed a reduced in thickness by approximately 60%.

Mechanical stimulation of the cells has been shown to affect the ability of cells to change their surrounding matrix.[44] The application of force onto cells can initiate several cellular processes including ion-channel activation, phosphorylation and cytoskeletal changes. Another factor that plays an important role in dictating the cell–matrix mechanical relationships is the initial conditions used to manufacture the hydrogels. Cell seeding density can also influence the mechanical properties of hydrogels. Increasing the cell number often leads to an increase in the rate of hydrogel remodelling and changes in mechanical properties. For fibroblast-seeded collagen hydrogels, the initial cell and collagen concentration used was found to affect the ability of cells to change the mechanical properties of the hydrogels.[43] Varying the

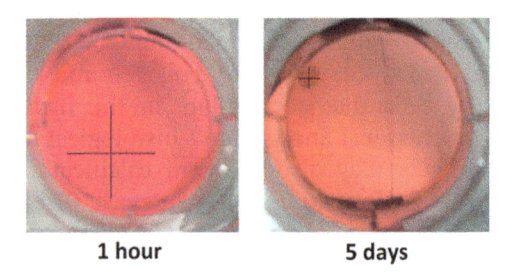

Figure 8.4 Images of a collagen hydrogel seeded with corneal fibroblasts that have undergone contraction.[13] Reproduced with permission from Mary Ann Liebert Inc., previously printed in *Tissue Engineering Part C*.

initial cell density in turn varies the amount of force that can be generated to remodel the hydrogels. This phenomenon demonstrates a clear mechano-feedback response between fibroblasts and their surrounding hydrogel matrix.

8.5 Mechanical Properties as a Marker of Cellular Activities

The mechanical properties of soft tissues are often closely related to their physiological function. For example, *in vivo* tissue contraction, remodelling and fibrosis (or scarring) following injury or disease often results in an alteration to the mechanical properties of the affected tissue[45] due to an 'activation' of the native cell phenotypes into their injury subtypes.[46] The dynamic reciprocity between hydrogel mechanical properties and cell activity has driven researchers to investigate how the mechanical properties of hydrogel constructs affects cell behaviour and whether these properties can act as a marker to predict cellular activity. This relationship has to be considered when designing hydrogels that need to be suitable for implantation. Here we present several examples to demonstrate how the mechanical properties of hydrogel constructs can reflect cellular activities.

8.5.1 Indicator of Differentiation Status

A recent study by Wilson *et al.*[14] demonstrated how the assessment of mechanical properties in terms of elastic modulus measurement could be used to determine the effect of biochemical ingredients and topographic features on corneal stromal cell differentiation in collagen hydrogels. The aim was to determine the most suitable culture condition whereby cultured corneal stromal cells that were initially 'activated' or fibroblastic in phenotype could be restored to the native 'inactivated' keratocyte phenotype *in vitro*. The basis of using matrix contraction capacity as an additional marker of corneal stromal cell phenotype differentiation in response to different culture conditions is that corneal keratocytes are quiescent and non-contractile and corneal fibroblasts are contractile and motile.[47] Thus, the construct contraction and elastic modulus measurements combined provided a descriptive insight into what was happening at a cellular level within the constructs when culture conditions change. This was used to indicate that the synergistic effect of nanofibre incorporation, serum removal, plus insulin medium supplementation provided the most suitable environment for the restoration of the native corneal keratocyte cell phenotype. These results were then corroborated with microscopic and genotypic characterization data to further validate that mechanical characterization can act as a sensitive marker of cellular activities such as cell phenotype and differentiation (Figure 8.5).

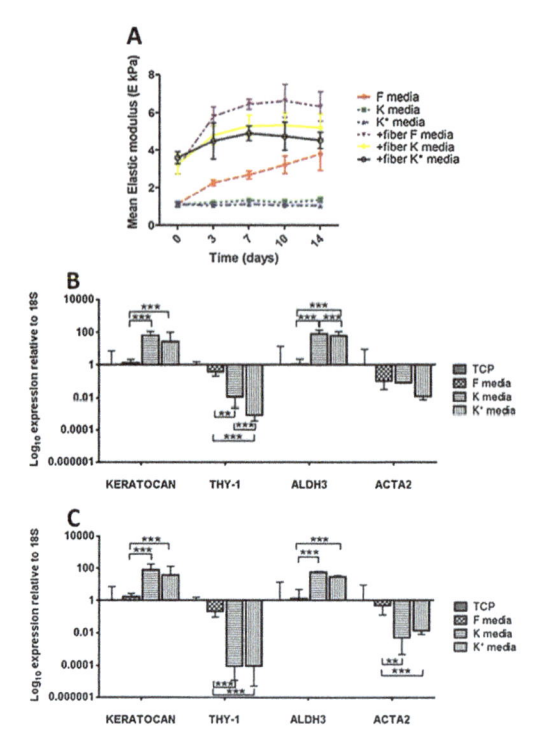

Figure 8.5 (A) The elastic modulus and gene expression of corneal stromal cells grown in collagen hydrogels for 14 days in response to chemical and topographic regulation. F denotes serum-containing medium, K denotes serum-free, insulin supplemented medium, and K′ denotes serum-free β-FGF supplemented medium; +fibre indicates the incorporation of nanofibres in the hydrogel. (B) The gene expression without nanofibre incorporation; (C) the gene expression with nanofibre incorporation; *keratocan* and *ALDH3* are keratocyte-specific genes, while *Thy-1* and *ACTA2* are corneal fibroblast-specific genes.[14] Reproduced with permission from John Wiley and Sons Ltd. Previously printed in *Advanced Functional Materials*.

8.5.2 Indicator of Cell Viability and Contractility

Ahearne *et al.*[13] used a spherical indentation technique to investigate the relationship between cell viability, hydrogel contraction and hydrogel elastic modulus in response to long-term culture (Figure 8.6). It was found that an initial increase in elastic modulus coincided with contraction of the hydrogel while a reduction in cell viability after several weeks in culture resulted in a reduced modulus. Inhibition of contraction using an MMP inhibitor found that when contraction was prevented, there was no subsequent increase in modulus. It was also found that the inhibition of actin stress fibres resulted in a reduction in elastic modulus, suggesting that the intrinsic strain applied by these cells was instrumental in controlling the bulk mechanical properties of the hydrogel. The actin staining images at corresponding time points

Mechanical Characterization of Hydrogels and its Implications for Cellular Activities 183

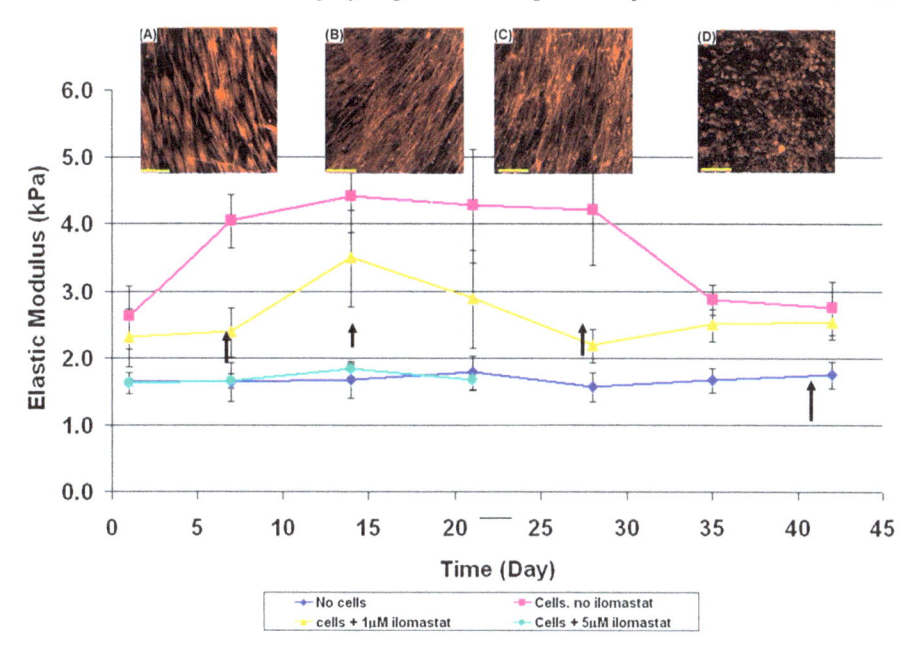

Figure 8.6 Change in elastic modulus of collagen hydrogels seeded with corneal stromal fibroblasts in response to culture time and the MMP inhibitor ilomastat.[13] Corresponding actin stained specimens at (A) 7 days; (B) 14 days; (C) 21 days and (D) 42 days. Reproduced with permission from Mary Ann Libert Inc., previously printed in *Tissue Engineering Part C*.

exhibited clear morphology difference in responding to the associated modulus. The specimens exhibited higher modulus expressed highly stretched actin filaments (Figure 8.6B and C), while the destroyed actin morphology (Figure 8.6D) appeared at the specimens which had low modulus with long culture duration, implying the low cell viability.

8.5.3 Indicator of Network Structure in the Hydrogel

There have been a large number of reports dedicated the effect of ageing on protein structures; in particular collagen type I, as it is a key lifelong structural protein in the body. A prevalent ageing mechanism, concerned with the non-enzymatic glycation of collagen, is the formation and accumulation of advanced glycation end-products (AGEs).[48] Accumulation of AGEs in relation to increasing chronological age has been linked to permanent alterations to the intra- and intermolecular structure of collagen, which often manifests as compromised mechanical properties to the tissue or construct being investigated. In recent work by Wilson *et al.*,[49] type I collagen was extracted from the tendons of different aged rats, varying from 2–3 days (newborn) to 2 years (old adults). The mechanical properties of the resulting reconstituted hydrogel constructs were then measured using an indentation technique.[50] It was found that in acellular hydrogel scaffolds that there was a clear visible trend

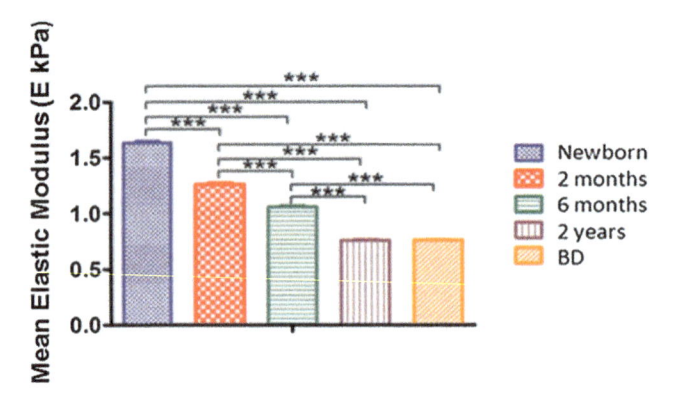

Figure 8.7 The elastic modulus of acellular hydrogel scaffolds using collagen extracted from rats of different ages.

showing that increasing age resulted in a reduction in the elastic modulus (Figure 8.7). The preliminary examination of the elastic modulus of corneal stromal fibroblasts grown in these hydrogels found that younger collagen induced higher contraction than older collagens manifesting as a higher modulus. Hence, it has been postulated that at a given collagen concentration, the younger collagen hydrogels (newborn and 2 months old) with a highly organized fibrous structure, resulted in a higher construct modulus compared to the randomly and loosely packaged older specimens (6 months and 2 years old). Thus, it is feasible to predict microscopic differences in the collagen hydrogel through the measurement of mechanical properties.

8.6 Strategies for Improving the Mechanical Properties of Hydrogels

When using hydrogels to study cell–ECM interactions, it becomes critical to tailor the hydrogels' mechanical properties. Various strategies have therefore been proposed to improve their mechanical characteristics. A fundamental limitation of hydrogels for tissue engineering is their inferior mechanical strength and stiffness in comparison to the native tissue that they are being used to replicate. These mechanical properties result from the high water content and random fibre orientations found in hydrogels.[7] Once the mechanical properties of a hydrogel material have been determined, it is often desirable to improve the mechanical strength of the construct so that it is more suitable for a given application.[2] The mechanical properties of hydrogels can be improved using numerous strategies including the alteration of the co-monomer composition, increasing/decreasing the cross-linking density, alterations to the conditions in which the polymer is formed,[2] the addition of cells onto or into the matrix *via* matrix remodelling, ECM secretions and the application of intrinsic strain.

8.6.1 Concentration

One approach to improving the mechanical properties of hydrogels is to increase the polymer concentration. Several studies have examined the relationship between mechanical properties and polymer concentration in hydrogels. Ahearne et al.[50] found that there was an almost linear increase in elastic modulus with hydrogel concentration when examining agarose and alginate hydrogels. Buckley et al.[37] found a similar trend when measuring the equilibrium and dynamic moduli of agarose hydrogels of increasing concentration. The elastic modulus of collagen hydrogels has also been shown to increase with concentration.[43] Interestingly, the initial collagen concentration also affected the subsequent rate of hydrogel contraction and matrix remodelling, with a lower initial collagen concentration having a faster rate of contraction. This faster rate of contraction led to these hydrogels having a higher cell density and a higher overall collagen density compared to the other hydrogels after 25 days in culture.[43]

Methods of increasing the concentration of hydrogels such as plastic compression have also demonstrated the relationship between hydrogel stiffness and hydrogel concentration.[51,52] By pushing fluid out of the hydrogels, this led to an increase in concentration thus an increase in stiffness. It has been reported that the polymer concentration in these hydrogels can increase by a factor of over 100.

8.6.2 Crosslinking

Chemical and photochemical crosslinking of matrix components such as collagen can also be used to influence the mechanical characteristics of hydrogels. Glutaraldehyde crosslinking of hydrogels has been shown to enhance the mechanical strength of several types of hydrogel.[53] The main problem with using glutaraldehyde is its toxicity. Alternative crosslinking agents such as genipin have been suggested as these are less toxic than glutaraldehyde.[53] UVA-crosslinking in the presence of riboflavin has been shown to increase the stiffness of collagen hydrogels without damage to the cells in those hydrogels.[15] UV light has also been used to develop hydrogels with a stiffness matrix gradient to allow for the study of hydrogel stiffness and cell behaviour.[54]

8.6.3 Composition

Altering the ratio of different monomers used to prepare a hydrogel is one of the simplest methods to increase the mechanical properties of the construct.[2] Provided that the hydrogel is not fabricated using identical monomer units, then by increasing the concentration of the physically stronger component, this should give a favourable outcome.

Alteration of the polymerization conditions can dramatically alter the final formed product.[2] Time, temperature and the amount and type of solvent used can all be altered accordingly. The volume of solvent used is of particular importance since it can alter crosslinking density, the type or nature of the solvent can alter the copolymer structure.[2] Post-polymerization techniques can also alter the network structure of a hydrogel, causing alterations to mechanical strength. In addition, thermal cycling of the polymer, which involves successive freezing and thawing cycles can also increase the mechanical properties of hydrogels.[2,55]

8.6.4 Orientation of Fibrous Components

Often, the native tissue architecture is pivotal to the *in vivo* mechanical strength and function of a tissue. Much research has focused upon the mimicking of native tissue architecture in both 2D and 3D cultures. Contact guidance techniques have been extensively researched as they affect several cell characteristics including orientation, morphology, differentiation and secretion of ECM proteins. It is the material composition and more specifically the 3D nano- and microscale structure (the mesostructure) of bio-artificial constructs that are pivotal to their success.[51] Micro- and nanopatterned surfaces, magnetic alignment and electrospinning techniques are among a variety of techniques utilized in order to achieve this.

8.6.5 Micro- and Nanopatterning

Micro- and nanopatterned surfaces are often manufactured by the use of templates with well-defined groove widths and depths into which cells with and without matrix materials are added.[56] The patterned surfaces effectively restrict random cell growth *via* the incorporation of either physical or biochemical barriers. Orientated deposition of ECM components is capable of reinforcing the substrate in a given direction, which enhances the global mechanical properties of the original construct.[56]

8.6.6 Magnetically Aligned Collagen

Magnetic fields have been utilized in an attempt to create orientated collagen type I fibrils.[57] The use of magnetic fields to induce collagen orientation is advantageous in that it is non-destructive.[57] It has been reported that molecules of collagen can be assembled into orientated fibrils *via* the application of a magnetic force.[57] In brief, this can be achieved by loading an aliquot of collagen into a shallow sample holder and positioning it horizontally in the central region of a split coil superconducting magnet and increasing the temperature from 20 to 30 °C for approximately 30 min. The collagen molecules assemble into orientated fibrils perpendicular to the applied field and transform into a viscous gel that is stable and orientated after the magnetic field is removed. A limitation of this technique is that fibril diameter cannot be regulated using this technique. Furthermore, there is conflicting evidence suggesting that the application of strong magnetic forces can in fact impair cell function and viability.[58]

8.6.7 Electrospinning of Nanofibres

Electrospinning is a process that is able to produce continuous fibres from the submicron down to the nanometre–diameter range.[59] These fibres can then be arranged to recreate the *in vivo* tissue microstructures and arrangements. Several studies have incorporated electrospun aligned nanofibres into hydrogels to improve mechanical properties and regulate cell behaviour. A schematic showing how aligned nanofibres meshes can be incorporated into a collagen hydrogel is shown in Figure 8.8. Wilson *et al.*[60] found that there was an increase in elastic modulus of collagen hydrogels seeded with corneal stromal cells after PLDLA nanofibres were added. The nanofibres also influenced the cell phenotype and cell orientation and reduced the rate of hydrogel contraction. Tonsomboon and Oyen[61] found a 10-fold increase in modulus after incorporating crosslinked gelatine nanofibres into alginate hydrogels. The combination of electrospun nanofibres and hydrogels represents an exciting new approach to engineering tissues with improved mechanical properties.

8.7 Conclusion

It has been demonstrated that the mechanical properties of hydrogels play a key role in the regulation of cellular activities and those cells are capable of remodelling the structural and mechanical properties of their surrounding hydrogel matrix. Understanding this reciprocal relationship is vital in the development of new tissue engineering and regenerative medicine strategies. It is envisioned that, by tailoring the mechanical characteristics of hydrogels to particular applications, more anatomically accurate tissues could be engineered.

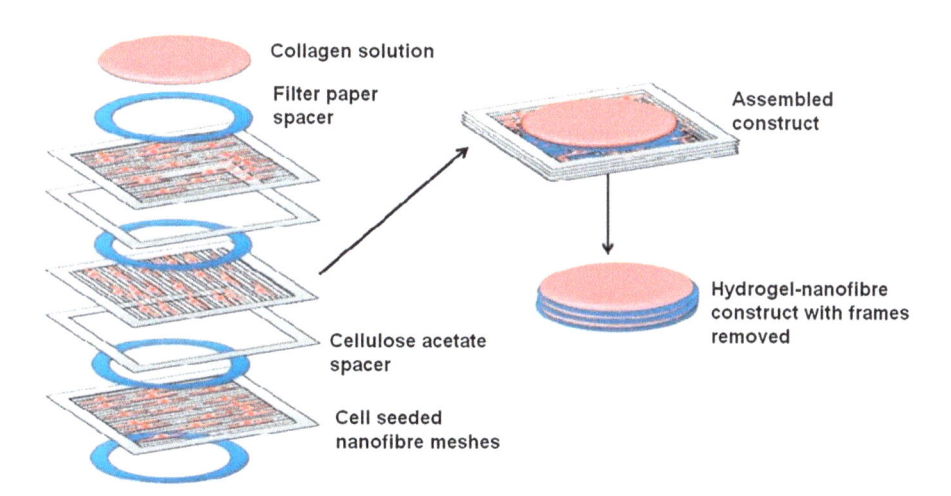

Figure 8.8 Schematic representation of the assembly process used to fabricate a nanofibre–hydrogel construct.[14] Reproduced with permission from John Wiley and Sons Ltd. Previously printed in *Advanced Functional Materials*.

References

1. A. M. Kloxin, C. J. Kloxin, C. N. Bowman and K. S. Anseth, *Adv. Mater.*, 2010, **22**, 3484–3494.
2. K. S. Anseth, C. N. Bowman and L. BrannonPeppas, *Biomaterials*, 1996, **17**, 1647–1657.
3. B. Mason, J. Califano and C. Reinhart-King, in *Engineering Biomaterials for Regenerative Medicine*, ed. S. K. Bhatia, Springer, New York, 2012, pp. 19–37.
4. J. D. Zieske, V. S. Mason, M. E. Wasson, S. F. Meunier, C. J. M. Nolte, N. Fukai, B. R. Olsen and N. L. Parenteau, *Exp. Cell Res.*, 1994, **214**, 621–633.
5. D. E. Discher, P. Janmey and Y.-L. Wang, *Science*, 2005, **310**, 1139–1143.
6. Y. Wang, G. Wang, X. Luo, J. Qiu and C. Tang, *Burns*, 2012, **38**, 414–420.
7. M. Ahearne, Y. Yang and K. K. Liu, *Topics in Tissue Engineering*, ed. N. Ashammakhi, R. L. Reis and F. Chielini, Kluwer, Dordrecht, 2008, pp. 1–16.
8. M. Ahearne and D. J. Kelly, *Biomed. Mater.*, 2013, **8**, 035004.
9. W. C. Hayes, L. M. Keer, G. Herrmann and L. F. Mockros, *J. Biomech.*, 1972, **5**, 541–551.
10. C. Pailler-Mattéi and H. Zahouani, *Tribol. Int.*, 2006, **39**, 12–21.
11. K. K. Liu and B. F. Ju, *J. Phys. D: Appl. Phys.*, 2001, **34**, 91–94.
12. B. F. Ju and K. K. Liu, *Mech. Mater.*, 2002, **34**, 485–491.
13. M. Ahearne, K. K. Liu, A. J. El Haj, K. Y. Then, S. Rauz and Y. Yang, *Tissue Eng., Part C*, 2010, **16**, 319–327.
14. S. L. Wilson, I. Wimpenny, M. Ahearne, S. Rauz, A. J. El Haj and Y. Yang, *Adv. Funct. Mater.*, 2012, **22**, 3641–3649.
15. M. Ahearne, Y. Yang, K. Y. Then and K. K. Liu, *Br. J. Ophthalmol.*, 2008, **92**, 268–271.
16. Y. Yang, P. O. Bagnaninchi, M. Ahearne, R. K. Wang and K. K. Liu, *J. R. Soc., Interface*, 2007, **4**, 1169–1173.
17. M. Ahearne, E. Siamantouras, Y. Yang and K. K. Liu, *J. R. Soc., Interface*, 2009, **6**, 471–478.
18. K. L. Johnson, *Contact Mechanics*, Cambridge University Press, Cambridge, Reprint edn, 1987.
19. J. D. Kaufman, G. J. Miller, E. F. Morgan and C. M. Klapperich, *J. Mater. Res.*, 2008, **23**, 1472–1481.
20. M. Eastwood, D. McGrouther and R. Brown, *Proc. Inst. Mech. Eng., Part H*, 1998, **212**, 85–92.
21. J. Fromageau, J.-L. Gennisson, C. Schmitt, R. L. Maurice, R. Mongrain and G. Cloutier, *IEEE Trans. Ultrason. Ferroelectrics Freq. Contr.*, 2007, **54**, 498–509.
22. C. Li, Z. Huang and R. K. Wang, *Opt. Express*, 2011, **19**, 10153–10163.
23. A. Samani and D. Plewes, *Phys. Med. Biol.*, 2004, **49**, 4395–4405.
24. J. Y. Rho, R. B. Ashman and C. H. Turner, *J. Biomech.*, 1993, **26**, 111–119.

25. B. Hinz, D. Mastrangelo, C. E. Iselin, C. Chaponnier and G. Gabbiani, *Am. J. Pathol.*, 2001, **159**, 1009–1020.
26. M. A. Griffin, S. Sen, H. L. Sweeney and D. E. Discher, *J. Cell Sci.*, 2004, **117**, 5855–5863.
27. T. Yeung, P. C. Georges, L. A. Flanagan, B. Marg, M. Ortiz, M. Funaki, N. Zahir, W. Ming, V. Weaver and P. A. Janmey, *Cell Motil. Cytoskeleton*, 2005, **60**, 24–34.
28. C. S. Chen, M. Mrksich, S. Huang, G. M. Whitesides and D. E. Ingber, *Science*, 1997, **276**, 1425–1428.
29. N. L. Halliday and J. J. Tomasek, *Exp. Cell Res.*, 1995, **217**, 109–117.
30. A. J. Engler, S. Sen, H. L. Sweeney and D. E. Discher, *Cell*, 2006, **126**, 677–689.
31. A. Banerjee, M. Arha, S. Choudhary, R. S. Ashton, S. R. Bhatia, D. V. Schaffer and R. S. Kane, *Biomaterials*, 2009, **30**, 4695–4699.
32. L. Bian, C. Hou, E. Tous, R. Rai, R. L. Mauck and J. A. Burdick, *Biomaterials*, 2013, **34**, 413–421.
33. A. J. Steward, D. R. Wagner and D. J. Kelly, *Eur. Cells Mater.*, 2013, **25**, 167–178.
34. B. Trappmann, J. E. Gautrot, J. T. Connelly, D. G. Strange, Y. Li, M. L. Oyen, M. A. Cohen Stuart, H. Boehm, B. Li, V. Vogel, J. P. Spatz, F. M. Watt and W. T. Huck, *Nat. Mater.*, 2012, **11**, 642–649.
35. A. J. Steward, S. D. Thorpe, T. Vinardell, C. T. Buckley, D. R. Wagner and D. J. Kelly, *Acta Biomater.*, 2012, **8**, 2153–2159.
36. C. Mauch, B. Adelmann-Grill, A. Hatamochi and T. Krieg, *FEBS Lett.*, 1989, **250**, 301–305.
37. C. T. Buckley, S. D. Thorpe, F. J. O'Brien, A. J. Robinson and D. J. Kelly, *J. Mech. Behav. Biomed. Mater.*, 2009, **2**, 512–521.
38. C. T. Buckley, T. Vinardell, S. D. Thorpe, M. G. Haugh, E. Jones, D. McGonagle and D. J. Kelly, *J. Biomech.*, 2010, **43**, 920–926.
39. J. C. Hu and K. A. Athanasiou, *Biomaterials*, 2005, **26**, 2001–2012.
40. T. E. Douglas, P. B. Messersmith, S. Chasan, A. G. Mikos, E. L. de Mulder, G. Dickson, D. Schaubroeck, L. Balcaen, F. Vanhaecke, P. Dubruel, J. A. Jansen and S. C. Leeuwenburgh, *Macromol. Biosci.*, 2012, **12**, 1077–1089.
41. B. M. Gillette, J. A. Jensen, B. Tang, G. J. Yang, A. Bazargan-Lari, M. Zhong and S. K. Sia, *Nat. Mater.*, 2008, **7**, 636–640.
42. E. Bell, B. Ivarsson and C. Merrill, *Proc. Natl. Acad. Sci. U. S. A.*, 1979, **76**, 1274–1278.
43. M. Ahearne, S. L. Wilson, K. K. Liu, S. Rauz, A. J. El Haj and Y. Yang, *Exp. Eye Res.*, 2010, **91**, 584–591.
44. V. S. Nirmalanandhan, N. Juncosa-Melvin, J. T. Shearn, G. P. Boivin, M. T. Galloway, C. Gooch, G. Bradica and D. L. Butler, *Tissue Eng.*, 2009, **15**, 2103–2111.
45. Y. Yang, P. O. Bagnaninchi, M. Ahearne, R. K. Wang and K.-K. Liu, *J. R. Soc., Interface*, 2007, **4**, 1169–1173.

46. M. E. Fini, *Prog. Retinal Eye Res.*, 1999, **18**, 529–551.
47. S. L. Wilson, A. J. El Haj and Y. Yang, *J. Funct. Biomater.*, 2012, **3**, 642–687.
48. J. M. Haus, J. A. Carrithers, S. W. Trappe and T. A. Trappe, *J. Appl. Phys.*, 2007, **103**, 2068–2076.
49. S. L. Wilson, M. Guilbert, J. Sulé-Suso, J. Torbet, P. Jeannesson, G. D. Sockalingum and Y. Yang, *FASEB J.*, 2014, **1**, 14–25.
50. M. Ahearne, Y. Yang, A. J. El Haj, K. Y. Then and K. K. Liu, *J. R. Soc., Interface*, 2005, **2**, 455–463.
51. R. A. Brown, M. Wiseman, C. B. Chuo, U. Cheema and S. N. Nazhat, *Adv. Funct. Mater.*, 2005, **15**, 1762–1770.
52. L. A. Micol, M. Ananta, E.-M. Engelhardt, V. C. Mudera, R. A. Brown, J. A. Hubbell and P. Frey, *Biomaterials*, 2011, **32**, 1543–1548.
53. J. V. Cauich-Rodriguez, S. Deb and R. Smith, *Biomaterials*, 1996, **17**, 2259–2264.
54. R. Sunyer, A. J. Jin, R. Nossal and D. L. Sackett, *PLoS One*, 2012, **7**, e46107.
55. O. Ariga, M. Kato, T. Sano, Y. Nakazawa and Y. Sano, *J. Ferment. Bioeng.*, 1993, **76**, 203–206.
56. N. E. Vrana, A. Elsheikh, N. Builles, O. Damour and V. Hasirci, *Biomaterials*, 2007, **28**, 4303–4310.
57. J. Torbet, M. Malbouyres, N. Builles, V. Justin, M. Roulet, O. Damour, A. Oldberg, F. Ruggieo and D. J. S. Hulmes, *Biomaterials*, 2007, **28**, 4268–4276.
58. O. Valiron, L. Peris, G. Rikken, A. Schweitzer, Y. Saoudi, C. Remy and D. Job, *J. Magn. Reson. Imag.*, 2005, **22**, 334–340.
59. W. E. Teo and S. Ramakrishna, *Nanotechnology*, 2006, **17**, 89–106.
60. S. L. Wilson, I. Wimpenny, M. Ahearne, S. Rauz, A. J. El Haj and Y. Yang, *Adv. Funct. Mater.*, 2012, **22**, 3641–3649.
61. K. Tonsomboon and M. L. Oyen, *J. Mech. Behav. Biomed. Mater.*, 2013, **21**, 185–194.

CHAPTER 9

Extracellular Matrix-like Hydrogels for Applications in Regenerative Medicine

ALEKSANDER SKARDAL

Wake Forest Institute for Regenerative Medicine, Wake Forest School of Medicine, Medical Center Boulevard, Winston-Salem, NC 27157, USA
E-mail: askardal@wakehealth.edu

9.1 A Brief Introduction to the Field of Biomaterials

Today the term 'biomaterial' encompasses a vast range of materials, including technologies that did not exist even a decade ago. Biomaterials range from cell supportive soft hydrogels, as we discuss in this chapter, to stiff metal or ceramic implants; from nanoparticles and quantum dots for drug delivery and imaging, to complex functioning medical devices such as left ventricular assist devices and artificial hearts. As proficiency in material science and biology continues to expand, so does the number of classifications of biomaterial types.[1,2]

Here we focus specifically on hydrogel biomaterials and their implementation in regenerative medicine applications such as cell therapy and tissue engineering. With the exception of bone and teeth, hydrogels allow for mimicry of the range of elastic modulus (E') values associated with the soft tissues of the body. Furthermore, processing techniques to generate sol–gel transitions can be designed to be non-cytotoxic, allowing simple encapsulation of cells. This is important as there is increasing movement from 2D to

3D cell and tissue culture in the fields of tissue engineering, regenerative medicine, and tissue modelling.[3] Hydrogels that support encapsulation procedures are infinitely more efficient for 3D uses than rigid scaffold seeding approaches of the past. While not every example discussed in this chapter reflects 3D use, the majority do, and it is important to note that in general, 3D applications in regenerative medicine provide cellular environments more like those in the body.

9.2 Hydrogel Biomaterial Types

The majority of hydrogel biomaterials typically fall into one of two major categories: synthetic hydrogels which are completely synthesized in the laboratory, or naturally derived hydrogels which are purified from natural sources and often further modified in the laboratory. Common examples of synthetic hydrogels include poly(ethylene glycol) (PEG)-based materials, such as PEG diacrylate (PEGDA), as well as polyacrylamide (PAAm)-based gels. Examples of naturally derived materials that are commonly used to generate hydrogels include collagen, hyaluronic acid (HA), alginate, and fibrin. In general, synthetic materials allow for more fine-tuned control over molecular weight numbers and distributions, as well as crosslinking densities, allowing for precise modulation of specific mechanical properties such as E' or stiffness. Natural hydrogels, on the other hand, often have an innate bioactivity that aids with cell and tissue integration and biocompatibility. In this chapter we briefly discuss the use of some common synthetic hydrogel biomaterials, but primarily focus on naturally derived hydrogel biomaterials, as they are more efficient at mimicking the biological nature of the native extracellular matrix (ECM).

9.2.1 Synthetic Polymer Hydrogels

A variety of synthetic materials have been implemented as hydrogels for applications in regenerative medicine. Synthetic polymers are advantageous for one primary reason—as indicated above, they allow for precise control over their chemical and physical properties. Scientists can maintain precise chemical control over molecular weight, functional groups, and hydrophobicity/hydrophilicity at a monomer level. As a result, crosslinking rates and mechanical properties can be precisely controlled. PEG and PAAm are examples of commonly used synthetic polymers in biomedical applications. PEG, which is perhaps most common, has long been used as medical device coatings to control host immune responses or appended to drug constructs to reduce degradation *in vivo*. It can also be manipulated to form a variety of hydrogels for cell culture and stem cell differentiation. PEG is often chemically modified with acrylate groups to create a photopolymerizable PEG diacrylate (PEGDA) in which cells can quickly be encapsulated.

The same features that allow such precise control over the chemical and mechanical properties also translate into an inherent drawback. Since synthetic polymer chains typically do not contain natural attachment sites that can interact with cells, all biological activity must be artificially pre-programmed into the material. PEG requires chemical immobilization of cell adhesion motifs in order to support cell adherence. Alternatively, many hydrogels derived from natural polymers and peptides retain some, if not all, of their original biological activity.

9.2.2 Collagen

Collagen is one of the most frequently used natural materials for a cell substrate, since it is the most abundant component of the ECM in most tissues.[4] Isolation and purification processes are well established, particularly for collagen type I, so using collagens as surface coatings and gels for cell culture has become an industry-wide practice. Inherent in the collagen structure are important arginine–glycine–aspartic acid (RGD) amino acid sequences that allow cells to adhere and proliferate *via* integrin–RGD binding. However, in normal tissue and ECM, collagen is but one of many components. Collagen biomaterial matrices are indeed useful and have yielded many important biological advances, but ~100% collagen matrices may limit cell migration and locomotion due to strong cell attachment. The lack of other common ECM components such as elastin, fibrinogen, laminin, and glycosaminoglycans, may result in biological signalling that can induce unanticipated cellular changes. Furthermore, collagen fibres and gels primarily contain hydrophobic peptide motifs. As such, when used as implants or cell delivery agents, collagen gels can exclude water and contract, potentially resulting in decreased function, decreased diffusion of nutrients and gases, and cell death. Despite this limitation, collagen is still used extensively in tissue culture. However its future application might be improved with development of new hybrid biomaterials consisting of combinations of collagen and other ECM components with superior properties.

9.2.3 Hyaluronic Acid

Hyaluronan (or HA), is a versatile non-sulfated glycosaminoglycan (GAG) consisting of repeating disaccharide units and is present in tissues as a major constituent of the ECM that has shown great potential in regenerative medicine.[5,6] Unmodified HA has been used clinically for over three decades,[7] in applications such as treatment of damaged joints.[8,9] More recently, HA has been commonly chemically modified to become a more useful and robust biomaterial that can be crosslinked or loaded with other functional molecules.[10]

HA hydrogels are often implemented by photocrosslinking methacrylate groups appended to the HA chains that can undergo free-radical polymerization when exposed to ultraviolet (UV) irradiation to form soft hydrogels, referred to here as MA–HA hydrogels. Photocrosslinkable MA–HA hydrogels have been used in many settings, from cutaneous and corneal wound healing[11] to prototype vessel structure bioprinting.[12] Thiol-modification of HA also yields a material by which hydrogels can be formed through Michael-type addition crosslinking. Like the MA–HA variety of HA, thiol-modified HA, particularly a thiolated carboxymethyl HA (CMHA-S), has been implemented in many applications in regenerative medicine such as wound healing,[13] tumour modelling,[14] and bioprinting of cellularized structures.[15]

9.2.4 Alginate

Alginate is a natural polysaccharide that is derived from algae or seaweed. It is commonly used in regenerative medicine applications because of the ease with which it can form a hydrogel through an almost instantaneous sodium–calcium ion exchange reaction. This has made alginate the material of choice for microencapsulation of cells, in which easily available and inexpensive alginic sodium salt (sodium alginate), which is unmodified, is quickly crosslinked into calcium alginate hydrogel microspheres.[16] These constructs have been extensively used for creating hydrogel capsules containing trapped liver cells or pancreatic islets.[17] However, without chemical modification, alginate, like PEG, is mostly inert, and use for cell and tissue culture is limited without incorporating cell-adherent motifs. Additionally, the reagents commonly used for creating cell-laden hydrogel microspheres, such as $CaCl_2$, the crosslinking reagent, as well as sodium citrate and ethylenediaminetetraacetic acid (EDTA), commonly used chelators, can have a detrimental effect on cell viability during the encapsulation process.[18] However, because of the ease with which alginate gels can be formed, it remains a popular and effective choice as a material in applications requiring cell encapsulation.

9.2.5 Fibrin

Another natural-sourced material for generating hydrogels is fibrin, which has been implemented for culture of various tissues types. Fibrin is made up of fibrinogen monomers that are joined by thrombin-mediated cleavage crosslinking. In the body, it has an important role in blood clotting, wound healing, and tumour growth. In a concentrated glue-like form, it has been used clinically as a haemostatic agent and sealant in surgery. More recently, less concentrated fibrin gels have been used as a scaffold for regenerative medicine due to their quick crosslinking rates and robust mechanical properties.[19]

9.3 Implementations in Regenerative Medicine

Regenerative medicine encompasses a wide range of subfields, research directions, and end applications. Here we discuss the implementation of the hydrogels mentioned above in applications such as culture of stem cells and primary cells, 3D tissue modelling, and bioprinting fabrication.

9.3.1 Stem Cell Culture

One application in which hydrogel biomaterials have become instrumental is attempting to mimic microenvironments of cells in ways that are simplified and deconstructed, in order to support and enable a targeted cell or tissue behaviour. Here we explore hydrogel biomaterial advances that have been implemented in stem cell biology or that have the potential to be useful in such applications. Specifically, we will discuss how hydrogels have been employed to drive differentiation of stem cells and generate artificial stem cell niches.

9.3.1.1 Driving Differentiation

The most comprehensively researched implementation of hydrogels with respect to stem cells is their use in directed differentiation. In this area of application much of the work has focused on differentiation towards osteogenic and chondrogenic lineages, although differentiation towards other lineages has also been explored. Hydrogels made up of PEG-based polymer backbones have been particularly widely used for chondrogenic differentiation. Bone marrow-derived mesenchymal stem cells (MSCs) were induced to undergo chondrogenesis in PEGDA gels *in vitro*[20] and *in vivo* in mice with a HA supplement.[21] As previously discussed, additional cell-adherent components are often added to PEG materials to provide support for cell attachment. For example, PEG–RGDS gels increased survival and chondrogenesis of MSCs, resulting in collagen, aggrecan, and GAG production like that of cartilage tissue.[22] By varying the composition of PEG hydrogels and addition of chondroitin sulfate, matrix metalloproteinase (MMP)-sensitive peptides, or HA, encapsulated MSCs could make more physiologically organized zone-specific cartilage-like tissue of varying compressive modulus.[23] PEG gels with RGD peptide motifs promoted human MSC viability by enhancing cell–matrix interactions, and induced mineralization by sequestering osteopontin with pendent phosphate groups.[24] Likewise, by adding phosphoester to PEG hydrogels, MSCs expressed bone-related markers and secreted osteocalcin, osteonectin, and alkaline phosphatase, resulting in increased mineralization and formation of bone-like tissue.[25] A thermoreversible and biodegradable Pluronic F127 and HA hydrogel was shown to have potential to drive adipose-derived stem cells (ADSCs) to undergo chondrogenesis.[26] Pluronic F127 hydrogels have also been implemented in adipogenic differentiation of MSCs when adipogenic factors were provided.[27] Dextran hydrogels crosslinked with four-armed PEG tetraacrylate linkers supported both chondrocytes and embryonic stem cells (ESCs) in culture for 3 weeks, after which both cell types produced cartilaginous tissue after incubation with chondrogenic media.[28]

Other work has focused on the impact of physical and spatial cues from the substrates on stem cells, rather than simply the material composition. In the following examples the precise modulation of the chemistries used for forming synthetic polymer materials allowed for targeted manipulation of the morphologies that the cells would experience in culture. Polyacrylamide gels were formed with topographical patterns made up of square and hexagonal posts of varying sizes and gap distances. The patterns were coated with collagen for cell attachment, after which seeded MSC morphologies substantially differed depending on the surface patterns.[29] A photomasking technique was used to pattern surface wrinkles in poly(2-hydroxyethyl methacrylate) surfaces in either lamellar or hexagonal patterns on which human MSCs were seeded. On lamellar patterns, MSCs differentiated into an osteogenic lineage, while on hexagonal patterns, MSCs differentiated into an adipogenic lineage.[30] Prominent in the literature is the work performed by Engler *et al.* in which it was demonstrated that the elastic modulus (E') of polyacrylamide gels coated with collagen determined lineage selection of MSCs. Substrates of 0.1–1 kPa induced a neuronal phenotype, substrates of 8–17 kPa induced a muscular phenotype, and stiff substrates of 25–40 kPa induced a osteogenic phenotype.[31] Further research has suggested that matching the E' of an *in vitro* matrix or substrate to that of the target tissue will improve and guide differentiation of the cells.

Recently, several HA-based gels have been implemented for culture of stem cells. Photocrosslinkable MA–HA hydrogels were used for manipulation of stem cell cultures, particularly controlled differentiation.[32] Three-dimensional cultures of human embryonic stem cells in MA–HA hydrogels were able to maintain an undifferentiated state for cell expansion, while differentiation could be selectively induced within the same hydrogel by exposure to the appropriate soluble factors.[33] In the same MA–HA hydrogels MSCs could successfully undergo chondrogenesis and showed increased collagen secretion rates in comparison to MSCs cultured in PEG hydrogels.[34] Furthermore, co-cultures of MSCs with articular chondrocytes increased the mechanical properties and improved ECM production in tissue engineered cartilage constructs.[35] MA–HA hydrogels have also been developed that support a double-crosslinking method used to create internal 3D patterns in hydrogels to spatially control encapsulated MSC spreading, which in turn could control adipogenic or osteogenic differentiation.[36] CMHA-S hydrogels formed from thiolated HA and gelatin have also shown promise in work with stem cells. Undifferentiated human ADSCs encapsulated in CMHA-S hydrogels underwent angiogenesis and adipogenesis showing feasibility of engineering viable adipose tissue.[37] CMHA-S hydrogels in combination with a perfusion bioreactor have also been used to support liver stem cell expansion and differentiation into hepatocytes.[38]

Fibrin has demonstrated usefulness in stem cell differentiation work also. Fibrin-only gels supported differentiation of bone marrow stromal cells into chondrocytes with higher levels of aggrecan and collagen II, and increased GAG deposition compared to fibrin alginate gels and standard cell pellet cultures.[39] A PEGylated fibrin gel was used to induce differentiation of ADSCs into endothelial and pericyte-like cells expressing CD31 and von Willebrand Factor (vWF), and used to form a vascularized dermal-like matrix.[40]

Endothelial progenitor cells (EPCs) seeded on fibrin gels showed increased cell viability as well as increased expression of Oct 3/4 and NANOG in comparison to parallel EPC cultures on fibronectin-coated plastic. When transitioned on to Matrigel for angiogenesis studies, the fibrin-cultured EPCs secreted higher levels of bioactive cell-recruitment cytokines.[41] PEGylated fibrin has been demonstrated to successfully support the formation of dental tissue. Cells derived from dental pulp and periodontal ligaments were combined with the fibrin gels to create constructs that were implanted in mice. These constructs had increased alkaline phosphatase levels, osteoblast gene expression, and dentin markers, and produced a collagenous matrix with mineral deposition.[42]

9.3.1.2 Niche Recapitulation

Hydrogel biomaterials are particularly relevant to stem cell research since they are inherently soft materials. The stiffness of body tissues ranges widely. For example, the elastic modulus of brain tissue is in the range of 0.5–2 kPa, muscle is near 10–15 kPa, while that of bone is more than 50 kPa. The stem cell niche is regarded by many as also being of a soft nature, and likely high in GAGs such as HA. Unfortunately, standard procedures call for culturing stem cells (as well as other cells) on simple plastic surfaces with a modulus that exceeds 10^5 kPa.[43]

Synthesis of biologically useful hydrogels with softer mechanical properties may provide tailorable microenvironments to act as cell niches, allowing scientists to move away from 2D settings towards more favourable 3D environments that increase the potential for physiological and clinical applications.[44] Indeed, more and more researchers are embracing the transition to 3D. Some hydrogels are formed in such ways that they can easily encapsulate stem cells for cell therapy delivery or *in vitro* engineering of tissues and organs. Hydrogels may also be useful for 3D stem cell expansion and recovery. Generally, recovery of cells, once encapsulated within a substrate, is difficult and requires harsh enzymatic or chemical treatment that can potentially impact cell viability and function. By combining CMHA-S materials with a cleavable disulfide bridge-containing PEGDA-like crosslinker, MSCs could be expanded with the benefits of 3D culture, and then recovered for future implantation or therapies using a mild enzyme-free treatment.[45] In other examples of expansion, spermatogonial stem cells showed increased cell attachment, viability, and proliferation in RGD-conjugated alginate hydrogels.[46] CMHA-S HA hydrogels loaded with hepatocyte growth factor (HGF) could be used for long-term recruitment of endogenous stem cells to injury or implant sites.[47]

Until recently, much work regarding mechanical properties of hydrogels and their impacts on stem cells has focused on how modulation of the elastic modulus of a substrate or microenvironment affects differentiation.[31] This was discussed above. However, pluripotency and the ability to differentiate into target cell types is not the only important use for stem cells: some stem cells are effective 'drug factories' possessing the ability to secrete biological compounds. In these cases, some biomaterials can help to maintain 'stemness' over time, *i.e.* protect the secretory ability of stem cells. Experiments in our laboratory have focused on determining the optimal substrate stiffness for the propagation and differentiation of perinatal stem cells from amniotic fluid and placenta. Immunohistochemical staining revealed increased surface expression of markers often associated with MSCs (CD44, CD90, CD105, and N-cadherin) in cells cultured on softer substrates, confirming that substrate stiffness alone may have profound effects on gene expression, and can thus alter the potential uses of these fetal stem cells.[48] In fact, growing these stem cells routinely on plastic in 2D culture could irreversibly alter their therapeutic potential.

As with MSCs, we have observed that, when in the right environment, perinatal stem cells are effective at releasing trophic factors that can induce angiogenesis and regeneration. Amniotic fluid stem (AFS) cells cultured on softer surfaces secreted factors that induced increased recruitment of other AFS cells and endothelial cells.[48,49] Additionally, preliminary results showed that conditioned media produced from AFS cells on soft substrates was more effective at inducing endothelial tube formation *in vitro* than conditioned media from AFS cells cultured on substrates with high elastic modulus including tissue culture plastic dishes. A similar phenomenon was noted when ESCs were cultured on poly-L-lysine/HA nanofilm of varying stiffness and chemistry. More highly crosslinked, and thus stiffer, films supported increased cell attachment and proliferation. However, on these stiffer films, expression of genes commonly expressed in the inner cell mass from where the ESCs were derived decreased. Only on softer, less-crosslinked films did the ESCs remain in colonies that maintained pluripotency.[50]

9.3.2 Primary Cell and Tissue Culture

Primary cells isolated from patient biopsies are generally difficult to maintain and expand efficiently using traditional culture methods *in vitro* without inducing notable changes in their biology. This is unfortunate as primary cells best reflect the behaviour and function of cells *in vivo*. Hydrogel materials may have an important role to play for improving culture conditions for these kinds of cells too.

Of particular interest to many researchers in the area of drug development are primary hepatocytes. Hepatocyte cultures are becoming increasingly popular as platforms for screening potential drug candidates and assessing drug metabolism and toxicity. Compared to hepatic tumour cell lines such as HEPG2 cells, primary hepatocytes are significantly more relevant for screening programmes, as they have *in vivo*-like rates of metabolism and proliferation, and are more susceptible to drug toxicity. However, simple cell-on-plastic culture approaches do not effectively support hepatocyte viability and function *in vitro*, necessitating alternative methods of culture. A number of studies address this need. For example, rat primary hepatocytes were encapsulated in barium alginate hydrogel fibres, which induced formation of hepatic cord structures. These structures remained viable for an impressive 90 days in culture, but hepatic function such as albumin and urea secretion, and gene expression, decreased after only 1–2 weeks.[51] In another approach, non-adherent PEG–fibrinogen hydrogels with no ECM components or growth factors (GFs) caused preservation of EGF and HGF receptors on rat primary hepatocytes. These 3D encapsulated cells had high levels of albumin and urea secretion for 10 days.[52]

These examples show promise, yet additional work needs to be undertaken to increase cell support for maintenance of function for longer periods of time. Additionally, while rat hepatocytes are undoubtedly useful, human hepatocyte cultures would be more reflective of drug interactions in actual human patients. To address these issues, we developed a HA-based hydrogel containing liver-specific ECM components and GFs, in which we preserved human primary hepatocyte viability and function for 4 weeks.[53] This study is further discussed later in the chapter.

Hydrogel technologies have been implemented to improve *in vitro* cultures of other primary cell types as well. Chondrocytes have extensively been biopsied and cultured in gels, with the goal being to generate replacement cartilage constructs. To investigate the effect of composition on cartilage formation, PEGDA was implemented in its normal form, or modified with proteoglycans, fibrinogen, or albumin. Articular chondrocytes were encapsulated in the gels and subjected to static of dynamic compression culture. Significant increases in GAG and collagen type II content in the PEG only and PEG–albumin constructs subjected to compression indicated that in the case of cartilage formation, a lack of cell–matrix interaction was actually important.[54] Peptide-based gels have also been used for primary chondrocyte culture. HLT2 peptides can spontaneously unfold and self-assemble to form gels in the appropriate conditions. Chondrocytes encapsulated in these gels stayed viable while secreting GAGs and collagen type II, resulting in a mechanically robust cartilage-like construct. Conversely, when a similar 'MAX8' peptide with increased charge density per monomer was used, chondrocytes died and formed poorly defined constructs.[55]

Several studies describe the use of alginate gels for the use of culturing harvest ovaries or ovarian follicles in order to generate viable oocytes. For example, co-cultures of murine ovarian follicles and mouse embryonic fibroblasts (MEFs) in alginate gels enabled *ex vivo* survival of the follicles for 14 days and production of oocytes. In gels without the MEFs, all follicles degenerated in less than 10 days.[56] With a fibrin-modified alginate gel, researchers were able to harvest follicles from *ex vivo*-cultured murine ovaries and encapsulate them in the gel. During a 12 day culture time course, these follicles produced oocytes in larger size and numbers than follicles cultured in alginate-only gels. Furthermore, oocytes from fibrin–alginate-cultured follicles could be successfully fertilized *in vitro*.[57] These advances have the potential for making an substantial impact in reproductive medicine.

9.3.3 Cell Therapy

Cell therapy is an area within regenerative medicine that has been prevalent for decades, but has encountered a number of hurdles standing in the way of reaching full clinical and commercial acceptance. Other than the obvious regulatory problems any cell-based product suffers, other difficulties include choosing the optimal delivery mechanism as well as poor cell engraftment and viability after administration. By combining cell therapy with biomaterials, one can deliver cells to a target area and improve engraftment through adhesion sites, and increase long-term cell viability at the site.[58] In many cell therapies, cells are simply delivered through syringes and needles. Unfortunately, viability of the therapeutic cell population is not always optimized, and it has been shown that delivering cells in a hydrogel can improve viability. In one example, researchers demonstrated that by encapsulating cells in alginate capsules, the viability of injected HUVECs, ADSCs, MSCs, and NPCs significantly increased cell survival in comparison to injections in liquid media alone.[59]

Fibrin gels have been valuable in cell therapy applications due to the quick crosslinking time supported by the enzymatic cleavage of fibrinogen by thrombin. Quick-gelling fibrin/collagen gels containing AFS cells were bioprinted over full-thickness skin wounds in nude mice to improve wound healing (Figure 9.1A–C). This treatment also increased vascularization (Figure 9.1D) and maturation of blood vessels (Figure 9.1E) in the regenerated tissue.[49] However, some healing occurred by wound contraction, which is less desirable than healing by re-epithelialization. We then evaluated a panel of 12 hydrogels for use in this *in situ* bioprinting approach for wound healing. The variety that performed the best in multiple categories was a photopolymerizable variety of the HA-based CMHA-S hydrogel that could be crosslinked *in situ* nearly instantaneously with a UV light.[60] This result led to an ongoing study in which we are pairing the beneficial properties of this HA hydrogel—anti-inflammatory, anti-scarring, associations with stem cell niches, non-swelling—with those of AFS cells in order to promote wound healing while minimizing fibrosis and scarring in regenerating skin. Preliminary results show that production of collagen type I, the primary component in scars, seems to be slowed down, while the presence of other important ECM components, such as GAGs, proteoglycans, and elastin, appear to be present in higher levels.

HA hydrogels have been used in many other cell therapy applications. Hydrogels consisting of HA and whole blood were implemented for stem cell transplantation in myocardial infarction. These gels are novel as they bind to the target tissue, increasing immobilization and chances for engraftment.[61] Similarly, cardiosphere-derived cells were delivered to myocardial infarct border zones in thiolated HA gels. These cells delivered *via* the hydrogel engrafted more efficiently and increased the left ventricle ejection fraction levels.[62] In an ocular application, thiol-crosslinked HA gels were injected *in situ* together with retinal progenitor cells in a photoreceptor injury model. Transplanted retinal progenitor cells (RPCs) had improved viability, distributed evenly in the sub-retinal space, and were positive for photoreceptor markers.[63]

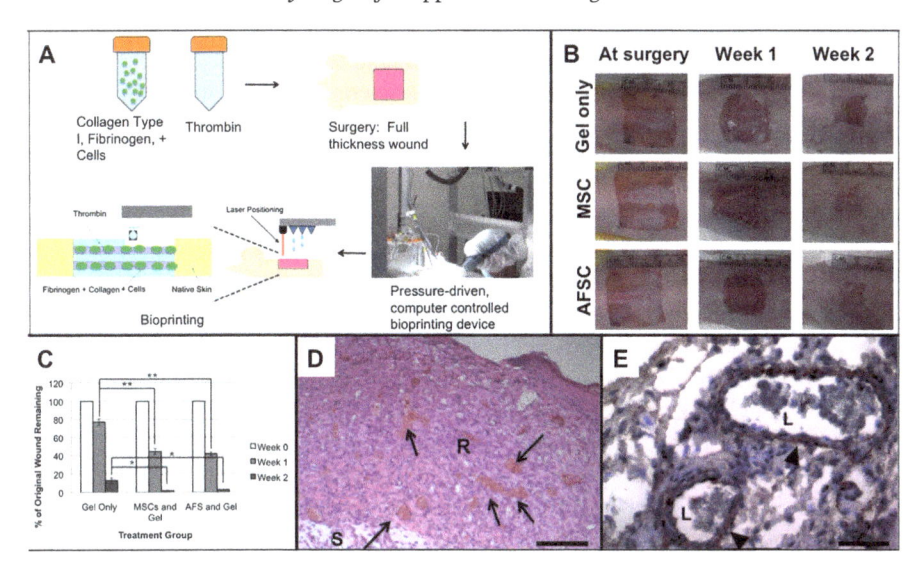

Figure 9.1 An example of amniotic fluid-derived stem (AFS) cells implemented as a cell therapy for wound healing by delivery in a fibrin–collagen gel (A). AFS cell delivery resulted in faster wound healing in mice, as shown by gross morphology (B), that was (C) significantly more effective than a gel-only treatment ($p < 0.05$). (D) AFS cell therapy resulting in an increased number of new blood vessels in the regenerating tissue that stabilized and matured, as demonstrated by alpha-smooth muscle actin staining (brown, indicated by arrows). Scale bar 50 μm. L, lumen; R, regenerated skin; S, subcutaneous tissue.

A variety of hydrogels were investigated (HA, collagen, tissue glue) to carry MSCs, intervertebral disc (IVD) cells, and chondrocytes for IVD therapy. It was found that the choice of material was extremely important to control the extent of cell proliferation, which in this case, could be detrimental to treating the injury.[64] In a rat acute kidney injury model, a thermoreversible chitosan hydrogel increased survival and engraftment of ADSCs, while also enhancing proliferation and reducing apoptosis of endogenous renal cells.[65] In an interesting example, researchers developed a pullulan hydrogel for MSC delivery. Pullulan is a carbohydrate produced by a fungus known for its anti-oxidant properties. Here, pullulan was shown to enhance MSC viability in both *in vitro* and *in vivo* environments of oxidative stress, commonly experienced at injury sites.[66]

9.3.4 Tissue Engineering *Ex Vivo/In Vitro*

Hydrogels have long been employed for use as scaffolds for engineering 3D tissues and organs. These laboratory-produced constructs may one day become commonplace for replacing diseased or damaged tissues in patients. Currently, they are being used as model systems in a variety of applications such as drug and toxicology screening.

9.3.4.1 3-D Modelling for Drug Screening

Researchers have built a vast biological knowledge base by mimicking how tissues behave *in vivo*, but in *in vitro* settings. To accomplish this effectively at low cost, one needs to implement small-scale models, but in large, high-throughput quantities. Many characteristics that are important in this area of application are similar to those discussed above regarding stem cell niche recapitulation and other applications. Namely, (1) 3D is superior to 2D, (2) cell–cell interactions and organization are important, and (3) the ECM or biomaterial choice must successfully support the tissue, whether it consists of primary cells, stem cell-derived cells, or a cell line. The closer to actual *in vivo* tissue a model system is, the more dependable and accurate it is in drug and toxicology screening.

A key aspect for accurately modelling pathologies and behaviour of tissues is the use of 3D systems since they allow cells to grow, differentiate, and interact with each other and the surrounding matrix, representing more *in vivo*-like conditions.[67,68] Although useful, 2D culture environments are not natural. For example, cancer drug testing using 2D cultures has shown some levels of success with candidate drugs, but when doses are scaled appropriately to patient levels, they are often ineffective. Conversely, 3D tumour models show an appropriate increased resistance to drugs, serving as better testing platforms.[69] This is likely due to the fact that *in vivo* tumour environments are 3D and much more complex in cell arrangement and tissue architecture. Furthermore, the diffusion properties of the drug are vastly different in 3D tissue in comparison to 2D monolayers.

Rotating wall vessel (RWV) technology is a cell culture system that has successfully been used to generate highly differentiated tissues *in vitro*.[70–72] The RWV is an optimized suspension culture system in which cells are grown in a physiological low-fluid-shear environment in 3D. Using this modality, cells are introduced in cylindrical bioreactors that are filled with culture medium and subsequently rotated along a horizontal axis. Continuous rotation of the bioreactors results in the gentle falling of cells through the medium while remaining in suspension, a state referred to as simulated microgravity (Figure 9.2A).[73] In the dynamic culture conditions of the RWV, cells naturally aggregate based on cellular affinities, form 3D structures and acquire properties of highly differentiated cells.[73–75] To date, more than 50 RWV-derived tissue models have been engineered, including liver, neuronal tissue, cardiac muscle, cartilage, adipose tissue, and epithelial tissues of the lung, bladder, small intestine, colon, placenta, and vagina.[76–87] Cells cultured in the RWV are able to recapitulate many of the fundamental aspects of parental tissue *in vivo*, including 3D spatial organization/polarity, cellular differentiation, multicellular complexity, and functionality. These 3D structures have also been validated in host–pathogen studies which demonstrated that the organotypic RWV-derived 3D cell models responded to bacterial and viral infection in ways that reflect the natural infection process *in vivo*—as evidenced by differences in tissue pathology, adherence and invasion,

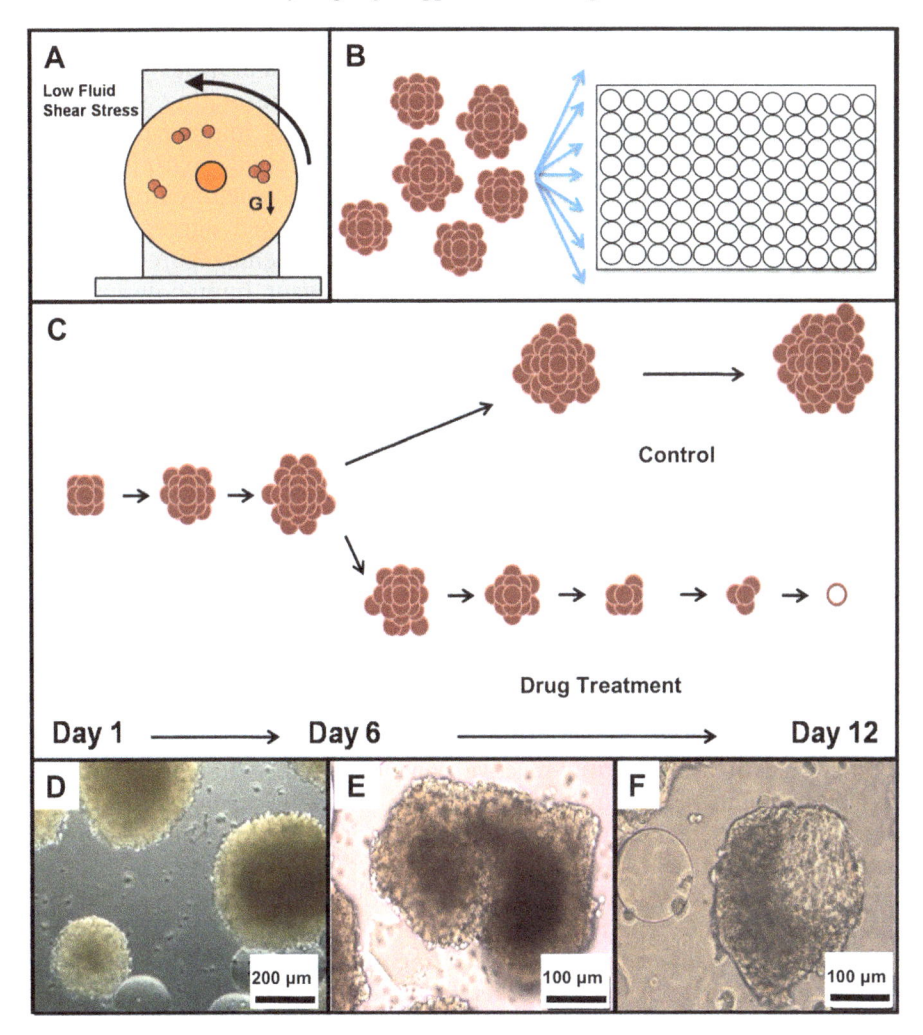

Extracellular Matrix-like Hydrogels for Applications in Regenerative Medicine 203

Figure 9.2 (A) Rotation at low rpm creates a low-fluid-shear stress environment in which cells and microcarriers are maintained in 'free fall' allowing cell aggregate or organoid formation. (B) After formation organoids are distributed into multiwell plates for high-throughput screening. (C) An example of a cancer drug screen performed on tumour organoids occurring in the above platform in which size is a metric for determining viability. (D–F) Examples of organoids formed by microcarrier–RWV culture of Int-407 intestinal epithelial cells, HepG2 C3A hepatoma cells, and MDA MB 231 breast cancer cells, respectively.

apoptosis and host biosignature profiles (proteomics and cytokine/chemokine profiling).[74,77,78,87–91]

Most of the above-described RWV-generated tissue models require the use of microcarrier beads for proper cell attachment and proliferation.[77–83,87] These microcarriers are typically submillimetre-sized beads made up of crosslinked polymers that provide sites for initial cell attachment. We developed a microcarrier bead coated with a previously designed HA and gelatin-based synthetic ECM (sECM) for use in rotating wall (RWV) vessel bioreactors.[92] The sECM hydrogel coating was integrated into the bead structure and provide a more compliant, hydrated environment for cell culture in the RWV bioreactor. Through RWV bioreactor culture, these beads gave rise to viable, well-differentiated, 3D cell aggregates. The generation of such organotypic tissues that closely resemble parental tissue *in vivo* provides affordable and physiologically relevant, human-derived test platforms for multiple disciplines such as infectious disease research, environmental toxicology, cancer research, and drug discovery. This technology is particularly suitable for high-throughput drug and toxicology screening protocols. Such an experimental approach is depicted in Figure 9.2B, where RWV-generated organoids are transferred to multiwell plates for testing of a large number of drugs at once. Figure 9.2C depicts one such screening outcome, where a drug induces cell death, decreasing the size of organoids. Figure 9.2D–F give examples of organoids formed by HA microcarrier-RWV culture of Int-407 intestinal epithelial cells, HepG2 C3A hepatoma cells, and MDA MB 231 breast cancer cells, respectively.

The HA–gelatin microcarriers described above were an iteration of a proven synthetic ECM hydrogel technology, consisting of CMHA-S and thiol-modified gelatin, which can be crosslinked *in situ* either by disulfide bonds or with bivalent thiol-reactive polymers such as PEGDA.[93,94] These sECM macromolecular building blocks were originally developed for use in regenerative medicine and tissue engineering.[95,96] They have proven exceedingly versatile with regard to the culturing of stem cells, primary cells, and cell lines *in vitro*, and have shown considerable promise in preclinical studies of wound repair,[97,98] adhesion prevention,[99,100] and tissue engineering,[101] including several 3D bioprinting applications.[12,15,102]

We developed a liver-specific version of this hydrogel for hepatocyte culture. To do this we incorporated a liver ECM-derived extract containing ECM components and GFs into an HA hydrogel. These hydrogels promoted primary human hepatocyte viability and function, including appropriate interconnected cellular morphology, albumin production, and the ability to metabolize an ethoxycoumarin-based drug for 4 weeks (Figure 9.3).[53] This result was significant compared to most examples in the literature in which rat hepatocytes rather than human hepatocytes were primarily used, and the cultures remained viable or functional for only 2–3 weeks. With this advance, we are currently applying simulated microgravity bioreactor culture with liver-specific hydrogel microcarriers and capsules to (1) generate hybrid liver tissue and cancer biopsy-derived tumour organoids for patient specific chemotherapy regiment optimization, and (2) develop microfabricated liver organoids to be used in a modular multi-tissue screening system for modelling the effects of infectious and biological agents for the military.

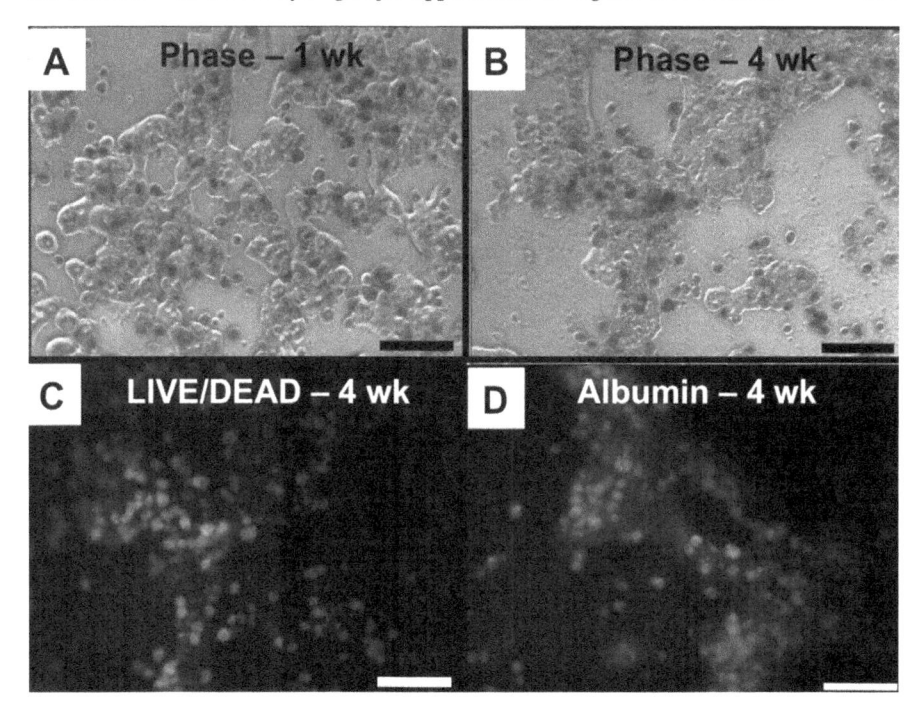

Figure 9.3 Liver-specific HA materials support hepatocyte viability and function for 4 weeks. Stable interconnected hepatocyte structures are present from (A) 1 week to (B) 4 weeks, as shown by phase-contrast microscopy. (C) At 4 weeks the majority of cells are viable. Green indicates calcein AM-stained live cells, while red indicates ethidium homodimer-stained dead cells. (D) Hepatocytes are actively producing albumin at 4 weeks, indicated by IHC staining. Significance $p < 0.05$. Scale bar 50 μm.

9.3.4.2 Bioprinting

To date, most engineered tissues consist of tissues of relatively simple structures or homogeneous cellular content, such as tubes, bone, cartilage, or layered patches, because of the difficulties associated with biofabricating complex, viable cellularized tissues. A common problem in creating tissue engineered constructs of significant size is supplying oxygen and nutrients throughout the entire construct. Without vasculature, the limits of diffusion through tissues limits oxygen transport to several hundred micrometres, placing a severe limit on the size of a construct that will remain viable in culture. To address this, some researchers attempt to prepattern 3D networks of lumens throughout the constructs to increase viability. This has been explored by biofabrication techniques such as bioprinting.

Bioprinting has emerged as a flexible tool with potential in a variety of tissue engineering and regenerative medicine applications. Most often it is implemented as an enabling technology for creating 3D tissue engineered constructs for either implantation or drug and toxicology screening, two areas described above. Bioprinting can be described as robotic deposition that has the potential to build organs or tissues.[103] In general, bioprinting uses a computer-controlled printing device to accurately deposit cells and biomaterials into precise 3D geometries in order to create anatomically correct structures. These devices have the ability to print cells (the 'bio-ink') in the form of cell aggregates, cells encapsulated in hydrogels, or cell-seeded microcarriers. The polymers that provide structure or space-holding capabilities are serving as the supporting 'bio-paper'.[104,105]

A number of printing techniques have been explored, each of which is dependent on the printable materials implemented. For example, printing techniques range from 'scaffold-free' printing (Figure 9.4A), to drop-by-drop patterning (Figure 9.4B), to continuous deposition (Figure 9.4C), and more. These approaches each have their own pros and cons.

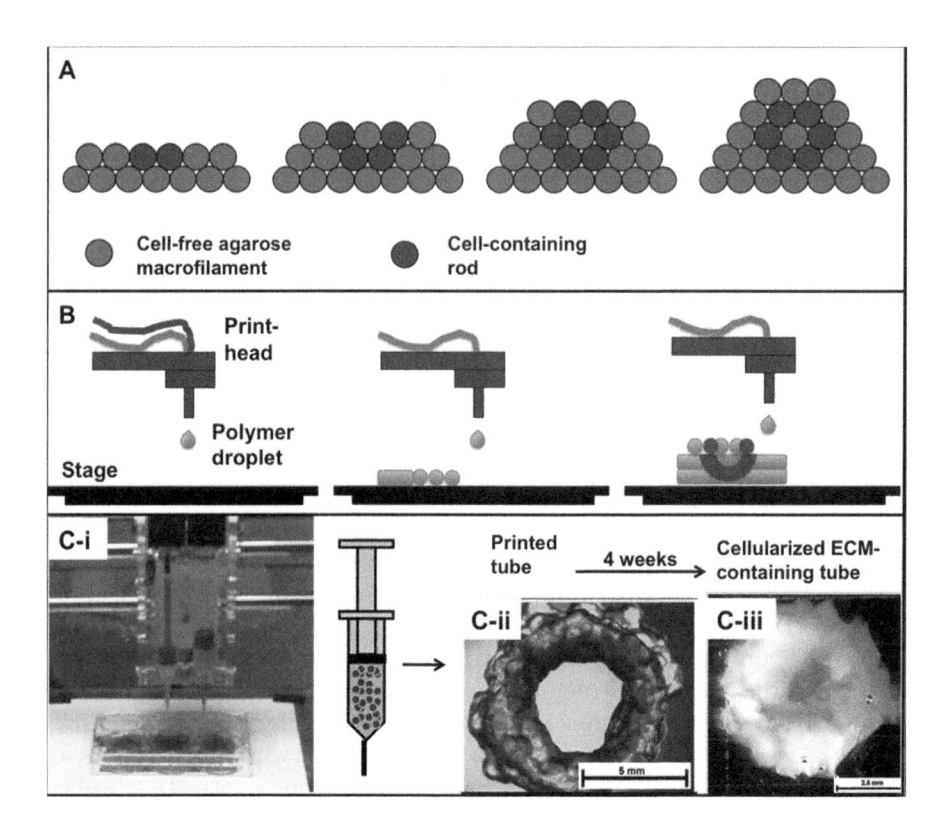

Figure 9.4 Three distinct bioprinting approaches: (A) 'scaffold-free' stacking of cell-free hydrogel macrofilaments and cell rods. (B) Drop-by-drop printing. (C) Continuous deposition from a syringe-based bioprinter, resulting in a tubular structure in which cells proliferated and secreted ECM over time in culture.

The 'scaffold-free' bioprinting technology is based on the principles of tissue liquidity and fusion. In this approach, aggregates or rods consisting solely of cells are printed in geometric patterns or shaped and allowed to fuse over time to form larger constructs.[106] In pioneering work, this approach was used to build branched vascular structures,[107] and more recently nerve grafts.[108] It should be noted that this type of bioprinting does actually rely on biomaterials, and as such the 'scaffold-free' term could be regarded as a misnomer. In most cases, the cell aggregates or cell rods are either printed into a biopolymer or additively stacked using space-holding biomaterials to preserve the appropriate structures during the tissue fusion and maturation process. Generally, these space-holders are eventually removed when the construct is sufficiently fused and possesses the mechanical properties to support itself. The strength of this method lies in its high cellularity, which allows for rapid fusion between discretely printed pieces. Unfortunately, preparation times, printing speeds, and building material volumes currently limit its scalability. Nevertheless, we explored a hydrogel-based approach that mimicked this technique. We developed a four-armed PEG tetraacrylate crosslinker to replace the linear PEGDA crosslinkers previously used with the HA gels we commonly worked with in order to yield hydrogels with increased elastic modulus. By increasing the stiffness of these materials, we were able to make HA hydrogel rods in which cells were encapsulated that could withstand the aforementioned stacking technique for creating tubular constructs *in vitro*.[102]

Drop-by-drop bioprinting, also often referred to as inkjet bioprinting, is another approach that is being explored for creating 3D biological structures, that is closely related to technologies used for cell patterning. Where basic cell patterning creates a 2D pattern made up of cells on a surface, by incorporating a hydrogel or other cell-friendly biomaterial, 3D cellularized structures can be fabricated drop by drop.[109,110] Examples of this implementation for 3D construct fabrication include collagen-encapsulated smooth muscle cells that were printed in droplet form to create muscular patches,[111] and the use of alginate and fibrin gel droplets for creating structures such as cellular fibre and multilayered cell sheets.[112] As discussed above, we have showed that this approach to bioprinting is also effective for skin printing to aid wound healing.[113] In terms of 3D fabrication, the drop-by-drop approach relies on being able to quickly polymerize the hydrogel in place, so that the next droplets can be added to the growing structure. Polymerization rates are a direct product of the various crosslinking chemistries innate to biomaterials. The requirement for a fast-gelling material places a limitation on the types of materials that can successfully be applied in this manner. Additionally, as the printable droplets are typically small volumes, scaling up to fabricate a large organ structure might be difficult. On the other hand, the small droplet volumes support high-resolution printing of intricate structures.

Finally, continuous deposition is alternative approach for 3D bioprinting that relies on the mechanical and temporal properties of the hydrogel being printed. Essentially, the problem becomes providing a material that is soft or fluid enough to facilitate extrusion through a syringe tip or nozzle, but that can support itself mechanically after deposition. This can be quite difficult when working with materials that rely on time for gelation to occur. Mistiming the deposition process can result in either a structure that collapses because crosslinking has not occurred quickly enough or, conversely, clogging of the bioprinting device as a result of polymerization occurring too rapidly. One example that addressed this problem was the combination of HA with dextran to form a semi-interpenetrating network-based gel that had the appropriate rheological properties for printing.[114] To address these issues we investigated various crosslinking techniques using HA-based hydrogels. First, we discovered that gold nanoparticles (AuNPs) could serve as thiophilic crosslinking agents when paired with thiolated HA and gelatin solutions. The gold–thiol interactions resulted in a hydrogel that gelled slowly, increasing in elastic modulus over the course of 96 h. This slow reaction produced a large window during which the material was extrudable for bioprinting (at about 24 h of crosslinking). We printed cellularized tubular structures that after layer-by-layer deposition, fused in culture during the next several days (Figure 9.4C). After 4 weeks in culture the constructs had become opaque with proliferating cells and cell-secreted ECM as they remodelled the construct. This crosslinking strategy was also reversible, allowing us to use cell-free AuNP gels as structural supports and space-holders that could be washed away by interrupting the gold–thiol bonds, resulting in a flexible system for building constructs.[15] We also explored the use of photocrosslinkable MA–HA and gelatin for continuous bioprinting deposition. This crosslinking strategy allowed an initial partial gelation step which left the gel in a soft and extrudable, but structurally sound, state during which cellularized tubular constructs were fabricated. After layer-by-layer deposition, the individual segments were fused and stiffened with a secondary photocrosslinking step. As in the previous example, these constructs were remodelled as the cells proliferated and deposited ECM.[12]

9.4 Future Potential and Implications

Often the novelty of a new hydrogel variety comes in the modifications that facilitate crosslinking strategies or new mechanical properties. Crosslinking strategies govern many aspects of application, as more applications are using the gelation process as part of a technique involving cells. Rather than preparing a hydrogel ahead of time as a substrate for cells to grow on top of, as researchers understand the benefits of 3D cultures, it is becoming more common to encapsulate cells within materials. The ease at which the crosslinking can be controlled determines how easily this can be done.

The particular mechanical properties of hydrogels are just as important. For example, there is a balancing act of tailoring the mechanical properties to a certain application, while maintaining the native properties of the base material after chemical modification. In the case of natural polymers, it is difficult to increase elastic modulus above a certain level. New approaches, be they crosslinking chemistries or combinations of materials, may be able to address this.

An area that is relatively new in biomaterial science is the development of hydrogels with the ability to dynamically change over time. This characteristic might be reflected in a slow stiffening or loosening over a long period of time, or a stimulus-activated fast change in stiffness. Work is currently progressing in this area, using materials that can be first crosslinked by one wavelength of light to form a cellularized gel, and then a light source at a second wavelength can be used to further stiffen or degrade localized regions of the gel to form 3D patterns of varying mechanical properties.[36,115]

Additionally, it has become clear that in the case of fabrication of large tissue constructs, such as those that would be necessary for whole organ replacement therapies, simple hydrogels may not be sufficient. Hydrogels are typically soft, resulting in difficulties in providing mechanical self-support in large constructs. Rather, hybrid approaches that combine the biological and cell-friendly properties of hydrogels with increased mechanical support of other materials such as stiff synthetic materials will likely be necessary. Essentially, stiff materials will act as skeleton-like scaffolding in which the hydrogel-based 'soft-tissue' can mature and become functional over time. Indeed, these kinds of hybrid tissue engineering approaches are already being explored by researchers in our laboratory and others. For example, bilayered constructs were fabricated using starch/polycaprolactone scaffolds and agarose hydrogels seeded with AFS cells to form constructs with distinct bone and cartilage regions, respectively.[116]

9.5 Conclusions

The success of engineered tissue constructs is inherently driven by both the biological characteristics and the chemical and mechanical properties of the implemented hydrogel materials. In most regenerative medicine modalities, we depend on the hydrogels to (1) provide an environment in which cells can thrive and the tissue can mature, (2) support crosslinking chemistries that allow for efficient deposition in complex geometries, and (3) be sufficiently mechanically stable to maintain those desired geometries over time. Further development of hydrogels exploring new chemistries and novel combinations and hybrids of existing materials has the potential to make a substantial impact in the continuously evolving areas of tissue engineering and regenerative medicine.

References

1. D. Williams, *Biomaterials*, 2011, **32**, 1–2.
2. D. F. Williams, *Biomaterials*, 2009, **30**, 5897–5909.
3. G. D. Prestwich, *Acc. Chem. Res.*, 2008, **41**, 139–148.

4. E. Hesse, T. E. Hefferan, J. E. Tarara, C. Haasper, R. Meller, C. Krettek, L. Lu and M. J. Yaszemski, *J. Biomed. Mater. Res., Part A,* 2010, **94**, 442–449.
5. D. D. Allison and K. J. Grande-Allen, *Tissue Eng.,* 2006, **12**, 2131–2140.
6. C. B. Knudson and W. Knudson, *Semin. Cell Dev. Biol.,* 2001, **12**, 69–78.
7. J. W. Kuo, *Practical Aspects of Hyaluronan Based Medical Products*, CRC/Taylor & Francis, Boca Raton, FL, 2006.
8. R. Galus, M. Antiszko and P. Wlodarski, *Pol. Merkuriusz Lek.,* 2006, **20**, 606–608.
9. A. Schiavinato, M. Finesso, R. Cortivo and G. Abatangelo, *Clin. Exp. Rheumatol.,* 2002, **20**, 445–454.
10. G. D. Prestwich and J. W. Kuo, *Curr. Pharm. Biotechnol.,* 2008, **9**, 242–245.
11. D. Miki, K. Dastgheib, T. Kim, A. Pfister-Serres, K. A. Smeds, M. Inoue, D. L. Hatchell and M. W. Grinstaff, *Cornea,* 2002, **21**, 393–399.
12. A. Skardal, J. Zhang, L. McCoard, X. Xu, S. Oottamasathien and G. D. Prestwich, *Tissue Eng., Part A,* 2010, **16**, 2675–2685.
13. K. R. Kirker, Y. Luo, S. E. Morris, J. Shelby and G. D. Prestwich, *J. Burn Care Rehabil.,* 2004, **25**, 276–286.
14. Y. Liu, X. Z. Shu and G. D. Prestwich, *Tissue Eng.,* 2007, **13**, 1091–1101.
15. A. Skardal, J. Zhang, L. McCoard, S. Oottamasathien and G. D. Prestwich, *Adv. Mater.,* 2010, **22**, 4736–4740.
16. E. Santos, J. Zarate, G. Orive, R. M. Hernandez and J. L. Pedraz, *Adv. Exp. Med. Biol.,* 2010, **670**, 5–21.
17. E. C. Opara, S. H. Mirmalek-Sani, O. Khanna, M. L. Moya and E. M. Brey, *J. Investig. Med.,* 2010, **58**, 831–837.
18. J. Cohen, K. L. Zaleski, G. Nourissat, T. P. Julien, M. A. Randolph and M. J. Yaremchuk, *J. Biomed. Mater. Res., Part A,* 2011, **96**, 93–99.
19. T. A. Ahmed, E. V. Dare and M. Hincke, *Tissue Eng., Part B,* 2008, **14**, 199–215.
20. C. G. Williams, T. K. Kim, A. Taboas, A. Malik, P. Manson and J. Elisseeff, *Tissue Eng.,* 2003, **9**, 679–688.
21. B. Sharma, C. G. Williams, M. Khan, P. Manson and J. H. Elisseeff, *Plast. Reconstr. Surg.,* 2007, **119**, 112–120.
22. C. N. Salinas, B. B. Cole, A. M. Kasko and K. S. Anseth, *Tissue Eng.,* 2007, **13**, 1025–1034.
23. L. H. Nguyen, A. K. Kudva, N. L. Guckert, K. D. Linse and K. Roy, *Biomaterials,* 2011, **32**, 1327–1338.
24. C. R. Nuttelman, M. C. Tripodi and K. S. Anseth, *Matrix Biol.,* 2005, **24**, 208–218.
25. D. A. Wang, C. G. Williams, F. Yang, N. Cher, H. Lee and J. H. Elisseeff, *Tissue Eng.,* 2005, **11**, 201–213.
26. H. H. Jung, K. Park and D. K. Han, *J. Controlled Release,* 2010, **147**, 84–91.
27. A. V. Vashi, E. Keramidaris, K. M. Abberton, W. A. Morrison, J. L. Wilson, A. J. O'Connor, J. J. Cooper-White and E. W. Thompson, *Biomaterials,* 2008, **29**, 573–579.

28. J. M. Jukes, L. J. van der Aa, C. Hiemstra, T. van Veen, P. J. Dijkstra, Z. Zhong, J. Feijen, C. A. van Blitterswijk and J. de Boer, *Tissue Eng., Part A,* 2010, **16**, 565–573.
29. M. J. Poellmann, P. A. Harrell, W. P. King and A. J. Wagoner Johnson, *Acta Biomater.,* 2010, **6**, 3514–3523.
30. M. Guvendiren and J. A. Burdick, *Biomaterials,* 2010, **31**, 6511–6518.
31. A. J. Engler, S. Sen, H. L. Sweeney and D. E. Discher, *Cell,* 2006, **126**, 677–689.
32. J. A. Burdick and G. Vunjak-Novakovic, *Tissue Eng., Part A,* 2009, **15**, 205–219.
33. S. Gerecht, J. A. Burdick, L. S. Ferreira, S. A. Townsend, R. Langer and G. Vunjak-Novakovic, *Proc. Natl. Acad. Sci. U. S. A.,* 2007, **104**, 11298–11303.
34. C. Chung and J. Burdick, *Tissue Eng., Part A,* 2009, **15**, 243–254.
35. L. Bian, D. Y. Zhai, R. L. Mauck and J. A. Burdick, *Tissue Eng., Part A,* 2011, **17**, 1137–1145.
36. S. Khetan and J. A. Burdick, *Biomaterials,* 2010, **31**, 8228–8234.
37. L. Flynn, G. D. Prestwich, J. L. Semple and K. A. Woodhouse, *J. Biomed. Mater. Res., Part A,* 2009, **89**, 929–941.
38. E. Schmelzer, F. Triolo, M. E. Turner, R. L. Thompson, K. Zeilinger, L. M. Reid, B. Gridelli and J. C. Gerlach, *Tissue Eng., Part A,* 2010, **16**, 2007–2016.
39. S. T. Ho, S. M. Cool, J. H. Hui and D. W. Hutmacher, *Biomaterials,* 2010, **31**, 38–47.
40. S. Natesan, G. Zhang, D. G. Baer, T. J. Walters, R. J. Christy and L. J. Suggs, *Tissue Eng., Part A,* 2011, **17**, 941–953.
41. M. C. Barsotti, A. Magera, C. Armani, F. Chiellini, F. Felice, D. Dinucci, A. M. Piras, A. Minnocci, R. Solaro, G. Soldani, A. Balbarini and R. Di Stefano, *Cell Proliferation,* 2011, **44**, 33–48.
42. K. M. Galler, A. C. Cavender, U. Koeklue, L. J. Suggs, G. Schmalz and R. N. D'Souza, *Regener. Med.,* 2011, **6**, 191–200.
43. J. L. Vanderhooft, M. Alcoutlabi, J. J. Magda and G. D. Prestwich, *Macromol. Biosci.,* 2009, **9**, 20–28.
44. C. R. Nuttelman, M. A. Rice, A. E. Rydholm, C. N. Salinas, D. N. Shah and K. S. Anseth, *Prog. Polym. Sci.,* 2008, **33**, 167–179.
45. J. Zhang, A. Skardal and G. D. Prestwich, *Biomaterials,* 2008, **29**, 4521–4531.
46. C. Chu, J. J. Schmidt, K. Carnes, Z. Zhang, H. J. Kong and M. C. Hofmann, *Tissue Eng., Part A,* 2009, **15**, 255–262.
47. J. Zhao, N. Zhang, G. D. Prestwich and X. Wen, *Macromol. Biosci.,* 2008, **8**, 836–842.
48. A. Skardal, D. Mack, A. Atala and S. Soker, *J. Mech. Behav. Biomed. Mater.,* 2013, **17**, 307–316.
49. A. Skardal, D. Mack, E. Kapetanovic, A. Atala, J. D. Jackson, J. Yoo and S. Soker, *Stem Cells Transl. Med.,* 2012, **1**, 792–802.

50. G. Blin, N. Lablack, M. Louis-Tisserand, C. Nicolas, C. Picart and M. Puceat, *Biomaterials,* 2010, **31**, 1742–1750.
51. M. Yamada, R. Utoh, K. Ohashi, K. Tatsumi, M. Yamato, T. Okano and M. Seki, *Biomaterials,* 2012, **33**, 8304–8315.
52. C. M. Williams, G. Mehta, S. R. Peyton, A. S. Zeiger, K. J. Van Vliet and L. G. Griffith, *Tissue Eng., Part A,* 2011, **17**, 1055–1068.
53. A. Skardal, L. Smith, S. Bharadwaj, A. Atala, S. Soker and Y. Zhang, *Biomaterials,* 2012, **33**, 4565–4575.
54. T. P. Appelman, J. Mizrahi, J. H. Elisseeff and D. Seliktar, *Biomaterials,* 2011, **32**, 1508–1516.
55. C. Sinthuvanich, L. A. Haines-Butterick, K. J. Nagy and J. P. Schneider, *Biomaterials,* 2012, **33**, 7478–7488.
56. D. Tagler, T. Tu, R. M. Smith, N. R. Anderson, C. M. Tingen, T. K. Woodruff and L. D. Shea, *Tissue Eng., Part A,* 2012, **18**, 1229–1238.
57. S. Y. Jin, L. Lei, A. Shikanov, L. D. Shea and T. K. Woodruff, *Fertil. Steril.,* 2010, **93**, 2633–2639.
58. M. M. Pakulska, B. G. Ballios and M. S. Shoichet, *Biomed. Mater.,* 2012, **7**, 024101.
59. B. A. Aguado, W. Mulyasasmita, J. Su, K. J. Lampe and S. C. Heilshorn, *Tissue Eng., Part A,* 2012, **18**, 806–815.
60. S. V. Murphy, A. Skardal and A. Atala, *J. Biomed. Mater. Res., Part A,* 2013, **101**, 272–284.
61. C. Y. Chang, A. T. Chan, P. A. Armstrong, H. C. Luo, T. Higuchi, I. A. Strehin, S. Vakrou, X. Lin, S. N. Brown, B. O'Rourke, T. P. Abraham, R. L. Wahl, C. J. Steenbergen, J. H. Elisseeff and M. R. Abraham, *Biomaterials,* 2012, **33**, 8026–8033.
62. K. Cheng, A. Blusztajn, D. Shen, T. S. Li, B. Sun, G. Galang, T. I. Zarembinski, G. D. Prestwich, E. Marban, R. R. Smith and L. Marban, *Biomaterials,* 2012, **33**, 5317–5324.
63. Y. Liu, R. Wang, T. I. Zarembinski, N. Doty, C. Jiang, C. Regatieri, X. Zhang and M. J. Young, *Tissue Eng., Part A,* 2012, **19**, 135–142.
64. H. Henriksson, M. Hagman, M. Horn, A. Lindahl and H. Brisby, *J. Tissue Eng. Regener. Med.,* 2011, **6**, 738–747.
65. J. Gao, R. Liu, J. Wu, Z. Liu, J. Li, J. Zhou, T. Hao, Y. Wang, Z. Du, C. Duan and C. Wang, *Biomaterials,* 2012, **33**, 3673–3681.
66. V. W. Wong, K. C. Rustad, J. P. Glotzbach, M. Sorkin, M. Inayathullah, M. R. Major, M. T. Longaker, J. Rajadas and G. C. Gurtner, *Macromol. Biosci.,* 2011, **11**, 1458–1466.
67. E. Lavik and R. Langer, *Appl. Microbiol. Biotechnol.,* 2004, **65**, 1–8.
68. J. L. Ifkovits and J. A. Burdick, *Tissue Eng.,* 2007, **13**, 2369–2385.
69. W. J. Ho, E. A. Pham, J. W. Kim, C. W. Ng, J. H. Kim, D. T. Kamei and B. M. Wu, *Cancer Sci.,* 2010, **101**, 2637–2643.
70. S. Navran, *Biotechnol. Annu. Rev.,* 2008, **14**, 275–296.
71. C. A. Nickerson, E. G. Richter and C. M. Ott, *J. Neuroimmune Pharmacol.,* 2007, **2**, 26–31.

72. R. P. Schwarz, T. J. Goodwin and D. A. Wolf, *J. Tissue Cult. Methods,* 1992, **14**, 51–57.
73. B. R. Unsworth and P. I. Lelkes, *Nat. Med.,* 1998, **4**, 901–907.
74. C. A. Nickerson and C. M. Ott, *ASM News,* 2004, **70**, 169–175.
75. J. Barrila, A. Radtke, S. Sarker, A. Crabbé, M. M. Herbst-Kralovetz, C. M. Ott and C. A. Nickerson, *Nat. Rev. Microbiol.,* 2010, **8**, 791–801.
76. B. Yoffe, G. J. Darlington, H. E. Soriano, B. Krishnan, D. Risin, N. R. Pellis and V. I. Khaoustov, *Adv. Space Res.,* 1999, **24**, 829–836.
77. A. J. Carterson, K. Honer zu Bentrup, C. M. Ott, M. S. Clarke, D. L. Pierson, C. R. Vanderburg, K. L. Buchanan, C. A. Nickerson and M. J. Schurr, *Infect. Immun.,* 2005, **73**, 1129–1140.
78. K. Honer zu Bentrup, R. Ramamurthy, C. M. Ott, K. Emami, M. Nelman-Gonzalez, J. W. Wilson, E. G. Richter, T. J. Goodwin, J. S. Alexander, D. L. Pierson, N. Pellis, K. L. Buchanan and C. A. Nickerson, *Microbes Infect.,* 2006, **8**, 1813–1825.
79. C. A. Nickerson, T. J. Goodwin, J. Terlonge, C. M. Ott, K. L. Buchanan, W. C. Uicker, K. Emami, C. L. LeBlanc, R. Ramamurthy, M. S. Clarke, C. R. Vanderburg, T. Hammond and D. L. Pierson, *Infect. Immun.,* 2001, **69**, 7106–7120.
80. B. Sainz, Jr, V. TenCate and S. L. Uprichard, *Virol. J.,* 2009, **6**, 103.
81. Y. C. Smith, K. K. Grande, S. B. Rasmussen and A. D. O'Brien, *Infect. Immun.,* 2006, **74**, 750–757.
82. T. A. Myers, C. A. Nickerson, D. Kaushal, C. M. Ott, K. Honer zu Bentrup, R. Ramamurthy, M. Nelman-Gonzalez, D. L. Pierson and M. T. Philipp, *J. Neurosci. Methods,* 2008, **174**, 31–41.
83. B. E. Hjelm, A. N. Berta, C. A. Nickerson, C. J. Arntzen and M. M. Herbst-Kralovetz, *Biol. Reprod.,* 2010, **82**, 617–627.
84. N. Bursac, Y. Loo, K. Leong and L. Tung, *Biochem. Biophys. Res. Commun.,* 2007, **361**, 847–853.
85. C. A. Frye and C. W. Patrick, *In Vitro Cell. Dev. Biol.: Anim.,* 2006, **42**, 109–114.
86. T. Yoshioka, H. Mishima, Y. Ohyabu, S. Sakai, H. Akaogi, T. Ishii, H. Kojima, J. Tanaka, N. Ochiai and T. Uemura, *J. Orthop. Res.,* 2007, **25**, 1291–1298.
87. H. L. LaMarca, C. M. Ott, K. Honer Zu Bentrup, C. L. Leblanc, D. L. Pierson, A. B. Nelson, A. B. Scandurro, G. S. Whitley, C. A. Nickerson and C. A. Morris, *Placenta,* 2005, **26**, 709–720.
88. K. M. Yamada and E. Cukierman, *Cell,* 2007, **130**, 601–610.
89. L. G. Griffith and M. A. Swartz, *Nat. Rev. Mol. Cell Biol.,* 2006, **7**, 211–224.
90. D. Kabelitz and R. Medzhitov, *Curr. Opin. Immunol.,* 2007, **19**, 1–3.
91. T. M. Straub, K. Honer zu Bentrup, P. Orosz-Coghlan, A. Dohnalkova, B. K. Mayer, R. A. Bartholomew, C. O. Valdez, C. J. Bruckner-Lea, C. P. Gerba, M. Abbaszadegan and C. A. Nickerson, *Emerging Infect. Dis.,* 2007, **13**, 396–403.

92. A. Skardal, S. F. Sarker, A. Crabbe, C. A. Nickerson and G. D. Prestwich, *Biomaterials,* 2010, **31**, 8426–8435.
93. G. D. Prestwich, in *Glycoforum,* 2001, http://glycoforum.gr.jp/science/hyaluronan/HA18/HA18E.html.
94. X. Z. Shu, Y. Liu, F. S. Palumbo, Y. Luo and G. D. Prestwich, *Biomaterials,* 2004, **25**, 1339–1348.
95. G. D. Prestwich, *J. Cell. Biochem.,* 2007, **101**, 1370–1383.
96. G. D. Prestwich, *Organogenesis,* 2008, **4**, 42–47.
97. Y. Liu, S. Cai, X. Z. Shu, J. Shelby and G. D. Prestwich, *Wound Repair Regen.,* 2007, **15**, 245–251.
98. R. R. Orlandi, X. Z. Shu, L. McGill, E. Petersen and G. D. Prestwich, *Laryngoscope,* 2007, **117**, 1288–1295.
99. Y. Liu, X. Z. Shu and G. D. Prestwich, *Fertil. Steril.,* 2007, **87**, 940–948.
100. Y. Liu, A. Skardal, X. Z. Shu and G. D. Prestwich, *J. Orthop. Res.,* 2008, **26**, 562–569.
101. Y. Liu, X. Z. Shu and G. D. Prestwich, *Tissue Eng.,* 2006, **12**, 3405–3416.
102. A. Skardal, J. Zhang and G. D. Prestwich, *Biomaterials,* 2010, **31**, 6173–6181.
103. R. P. Visconti, V. Kasyanov, C. Gentile, J. Zhang, R. R. Markwald and V. Mironov, *Expert Opin. Biol. Ther.,* 2010, **10**, 409–420.
104. N. E. Fedorovich, J. Alblas, J. R. de Wijn, W. E. Hennink, A. J. Verbout and W. J. Dhert, *Tissue Eng.,* 2007, **13**, 1905–1925.
105. V. Mironov, T. Boland, T. Trusk, G. Forgacs and R. R. Markwald, *Trends Biotechnol.,* 2003, **21**, 157–161.
106. K. Jakab, C. Norotte, B. Damon, F. Marga, A. Neagu, C. L. Besch-Williford, A. Kachurin, K. H. Church, H. Park, V. Mironov, R. Markwald, G. Vunjak-Novakovic and G. Forgacs, *Tissue Eng., Part A,* 2008, **14**, 413–421.
107. C. Norotte, F. S. Marga, L. E. Niklason and G. Forgacs, *Biomaterials,* 2009, **30**, 5910–5917.
108. F. Marga, K. Jakab, C. Khatiwala, B. Shepherd, S. Dorfman, B. Hubbard, S. Colbert and F. Gabor, *Biofabrication,* 2012, **4**, 022001.
109. S. Catros, J. C. Fricain, B. Guillotin, B. Pippenger, R. Bareille, M. Remy, E. Lebraud, B. Desbat, J. Amedee and F. Guillemot, *Biofabrication,* 2011, **3**, 025001.
110. B. Guillotin and F. Guillemot, *Trends Biotechnol.,* 2011, **29**, 183–190.
111. S. Moon, S. K. Hasan, Y. S. Song, F. Xu, H. O. Keles, F. Manzur, S. Mikkilineni, J. W. Hong, J. Nagatomi, E. Haeggstrom, A. Khademhosseini and U. Demirci, *Tissue Eng., Part C,* 2010, **16**, 157–166.
112. M. Nakamura, S. Iwanaga, C. Henmi, K. Arai and Y. Nishiyama, *Biofabrication,* 2010, **2**, 014110.
113. A. Skardal, D. Mack, E. Kapetanovic, A. Atala, J. D. Jackson, J. J. Yoo and S. Soker, *Stem Cells Transl. Med.,* 2012, **1**, 792–802.

114. L. Pescosolido, W. Schuurman, J. Malda, P. Matricardi, F. Alhaique, T. Coviello, P. R. van Weeren, W. J. Dhert, W. E. Hennink and T. Vermonden, *Biomacromolecules,* 2011, **12**, 1831–1838.
115. A. M. Kloxin, A. M. Kasko, C. N. Salinas and K. S. Anseth, *Science,* 2009, **324**, 59–63.
116. M. T. Rodrigues, S. J. Lee, M. E. Gomes, R. L. Reis, A. Atala and J. J. Yoo, *Acta Biomater.,* 2012, **8**, 2795–2806.

Subject Index

Abbreviations: ECM, extracellular matrix

acid–gel formation, alginates 99
actin 182–3
    filaments 85, 86, 87, 178, 183
adhesion
    cell, and other surfaces 177, 178
        motifs 83
    focal (intracellular sites) 86–7
adipose-derived stem cells (ADSCs) 200, 201
    directed differentiation 195, 196
adult stem cells 74
    in corneal regeneration 141–2
    see also specific types
advanced glycation end-products (AGEs) 183
ageing and protein structure 183–4
algae species for alginate isolation 135
alginate(s) 135–41, 194
    chemistry and structure 97–8, 140–1
    extraction/isolation and purification/production 99–103, 136–7
    gelation 101–2, 137–9
alginate hydrogels 95–111, 135–70, 194
    biophysical and biochemical properties influencing cell phenotype 151–9
    as cell culture scaffolds 141–51
    microcapsules/microspheres *see* encapsulated/microencapsulated alginate hydrogels
    modification (chemical) 106–9, 139, 155–9
    ovarian tissue culture 199
    production/gelation 101–2, 137–9
    structure 140–1
alkaline phosphatase 126, 128, 179, 195, 197
    in peptide self-assembly 117
allogeneic transplantation
    adult stem cells 74
    parathyroid and pancreatic tissue 159
amino acids
    conjugated with nucleobases and saccharides 41
    in design of peptide-based supramolecular hydrogels 116
amniotic fluid stem cells 198
amniotic membrane, human (HAM) 76, 77
amphiphiles, peptide 79–80
angiogenesis, therapeutic 149–50
animal models, hydrogelator biocompatibility 42–3
arginine–glycine–aspartic acid sequence *see* RGD sequence
arginine–glycine–aspartic acid–serine sequence *see* RGDS

Subject Index

aromatic motif–pentapeptide conjugates 33–6
aromatic peptide amphiphiles 116, 121, 126
articular cartilage defect repair 146, 161
attraction (electrostatic), ECM–solute 13
autologous adult stem cell transplantation 74

bacteria (bacterial cell)
    enzymatic hydrogelation in control of 126
    probiotic *see* probiotic bacteria
Bifidobacterium 96
    microencapsulated in alginate gels 105, 106
biocatalytic assembly of supramolecular hydrogels 115–24
biocompatibility of small peptide-based hydrogelators 31–47
bioinspired materials 78
biomaterials 73–94, 191–4
    instructive role 81, 82
    natural/biological 76, 77–8, 192, 193–4
        limitations 78
    next-generation, enzymes in fabrication of 113–14
    smart 79, 88, 114–15
    synthetic 76, 192, 192–3
    in tissue engineering 78–88
biomimics (biological mimics) 77, 78, 129, 141
    ECM *see* extracellular matrix
    nucleus pulposus 155
    regenerative medicine 88
bioprinting/tissue printing 149, 205–8
biosensing 126–8
blending technique with alginate microcapsules 109

blood, endothelial progenitor cell capture from 150
blood vessel formation, therapeutic induction 149–50
bone engineering 144–8, 161
bone marrow stem cells, bovine 159
bovine bone marrow stem cells 159
bovine serum albumin hydrogel 7
bulge test 174

calcium in alginate crosslinking 98, 137, 139
cancellous (trabecular) bone engineering 144–8
cancer cell lines, hydrogelator biocompatibility 37–8
carboxylate residues, alginates 98–9
cardiac patch 148–9, 163
cardiac progenitor cells 149, 157, 163
cartilage
    defect repair 146, 147, 161
    engineering 144–8
    formation (chondrogenesis) with mesenchymal stem cells 145, 159, 177–8, 195, 196
catalytic assembly of supramolecular hydrogels 115–24
cell(s)
    adhesion *see* adhesion
    culture *see* culture
    delivery of recombinant protein hydrogels to 48–72
    differentiation *see* differentiation
    fate, enzyme-responsive hydrogels in control and directing of 125–6
    imaging in 126–8
    mechanical properties of hydrogels influenced by activity of 178–81

cell(s) (*continued*)
    mechanical properties of hydrogels influencing 174–6, 181–4
    phenotype *see* phenotype
cell-based therapy (in general) 73–94
    clinical applications 75–6
    ECM-like hydrogels 200–1
    pharmaceutical industry involvement 75–6
charge interactions, ECM–solute 13
chitosan–alginate composite gel scaffold 144
chitosan-coated alginate microcapsules 107
    limbal epithelial cell cultivation 142
chondrocytes (and alginate gels) 146–8
    biochemical properties of gels and 158
    biophysical properties of gels and 154
    preclinical studies of therapeutic cell delivery 161
chondrogenesis with mesenchymal stem cells 145, 159, 177–8, 195, 196
collagen 61–4, 77–8, 180, 183–4, 185, 193
    ageing effects 183–4
    magnetically-aligned 186
    as natural biomaterial 77–8
collagen hydrogels 62, 142, 175, 176, 177, 178, 180, 181, 184, 185, 187, 193
    compressed 78
    fibrin and 200
collagen-like protein 62–4
combinatorial mutagenesis 56
composition of hydrogels, alterations improving mechanical properties 185–6
compression testing 174
computational algorithm-aided design in protein engineering 56
concentration, polymer, improving 185
conjugation
    glutamate–elastin-like protein 60
    small peptides with synthetic molecules 32, 33–6, 38–41
continuous deposition bioprinting 207–8
contraction, hydrogel 180
    as indicator of cell viability and contractility 182
controlled degradation of matrices 51–2, 113, 115
controlled drug release 128
convection (convective flow) 24–6, 27
cornea
    regeneration 141–2
    stromal cells (incl. fibroblasts)
        ageing and 184
        differentiation in collagen hydrogels 181
covalent crosslinked recombinant proteins 57–8, 59, 60, 61
craniofacial injury repair 147
crosslinking (in hydrogels) 185
    alginates 97, 98, 139
        calcium and 98, 137, 139
    in improvement of mechanical properties 185
    recombinant protein hydrogels 53, 55, 56, 57–65
    steady-state diffusion in hydrogels and 6, 7, 20–3, 27
cryopreservation, stem cell 76, 143–4
culture (cell/tissue) 141–51, 195–8, 198–9
    3D *see* three-dimensional cell culture
culture (cell/tissue), *see also in vitro*
    alginate hydrogels as scaffolds for 141–51

Subject Index

hydrogelator biocompatibility in cancer cell lines 37–8
primary, ECM-like hydrogels and 198–9
stem cell 195–8
cytochalasin in collagen gels 178
cytotoxicity of hydrogelators 33, 34, 35, 36, 37, 38, 41

decapentaplegic (Dpp) gradient formation in *Drosophila* 19–20
Descemet's membrane, corneal endothelial cells transplanted onto 142
differentiation
    elastic modulus and 181, 197–8
    mechanical properties of hydrogels indicating status 181
    stem cell
        directed 195–7
        mesenchymal *see* mesenchymal stem cells
diffusion, *see also* reaction–diffusion model
    Fick's first law of 3, 5, 9, 10
    Fick's second law of 3, 17, 24
    steady-state 2–8
        in ECM 2–5
        in hydrogels 5–6
dipeptides
    Fmoc 117, 118, 120, 122
    N-terminus unsubstituted, biocompatibility 36–8
diphenylalanine (conjugated to peptides)
    and D-glucosamine and nucleobases 40
    N-terminated 37
domain-level engineering in recombinant hydrogels 52–5
Dpp (decapentaplegic) gradient formation in *Drosophila* 19–20
drop-by-drop bioprinting 207

Drosophila
    decapentaplegic (Dpp) gradient formation 19–20
    resilin-like proteins sourced from 63
drugs, *see also* pharmaceutical industry
    controlled release 128
    development and screening 3D modelling for 202–4
    hepatocyte cultures and 198–9, 204
dynamic combinatorial library, enzyme-driven 120

Ect1/E6E7 cell line, hydrogelator biocompatibility 37–8
elastic modulus 140–1
    ageing and 184
    calculation 175, 181
    cell viability and contractility and 182
    differentiation and 181, 197–8
    hydrogel concentration and 185
    stem cell cultures and 197, 198
elasticity, substrate, cell response to 81–3, *see also* viscoelastic properties
elastin 58, 60, 61
elastin-like protein (ELP) 58–61
elastography, ultrasound 176
electrospinning of nanofibres 187
electrostatic attraction or repulsion, ECM–solute 13
embryonic stem cells (ESCs) 74, 142–4, 198
    alginate gels and 142–4
    biochemical properties of 156
    preclinical therapeutic studies 160
    niche recapitulation and 198
emulsion method of alginate hydrogels production 100, 101

encapsulated/microencapsulated alginate hydrogels (microcapsules) 103–9, 140, 142, 146, 157, 160, 194
  corneal regeneration and 141–2
  mesenchymal stem cells in 146
  modifications 106–9
  probiotic bacteria and 103–10
endothelial cells
  corneal, transplantation onto Descemet's membrane 142
  progenitor 149–50, 197
    biochemical properties of alginate gels and 157
    preclinical studies of therapeutic cell delivery 162
engineered protein hydrogels 48–72
enzymatic cleavage/degradation 21–2
  collagen-like proteins 62
enzyme-responsive hydrogels 112–34
  biomedical applications 124–9
epithelial cells, limbal 142, 152, 154
Escherichia coli
  enzymatic hydrogelation in control of 126
  streptococcal collagen-like protein 63
esterases in peptide self-assembly 117, 118
  control and directing of cell fate 125–6
ex situ sol to gel transition 56
ex vivo tissue engineering 201–9
excipients mixed with alginate microcapsules 107, 108, 109
extensiometry, strip 173–4
external gelation of alginate 101–2

extracellular matrix (ECM) 1–30, 50–2
  interactions 13–17, 50
    with ECM components 13–14, 178–80
    with hydrogel components 14–17
  mechanical stimulation of cells affecting their ability to change surrounding matrix 180
  soluble molecule transport/binding etc. in 50–2
    compared with hydrogels 1–30
  synthetic/mimics 58, 191–215
  structuring 83–4
extracellular proteins and cell delivery of recombinant proteins 50–2
extrusion method of alginate hydrogels production 99–100, 101
eye, alginate gels in therapeutic techniques 142

fibrin–collagen gels 200
fibrin-only gels 194
  cell therapy 200
  directed differentiation 196–7
fibrous components, orientation 186
Fick's first law of diffusion 3, 5, 9, 10
Fick's second law of diffusion 3, 17, 24
fluorene (F)-conjugated gelator peptide 34, 35
9-fluorenylmethyloxycarbonyl (Fmoc) groups 80, 117, 118, 120, 122–4, 128
fluorescence recovery after photobleaching 19–20
fluorescent intracellular imaging 126–7
Fmoc groups 80, 117, 118, 120, 122–4, 128

Subject Index

14-3-3 proteins 88–9
free energy 118, 119
    Gibbs 23
functionality (and functionalized regions) of hydrogels 13, 15–17, 113–14
    Fmoc dipeptides and 122–3

GAGAS pentapeptide–aromatic motif conjugates 34, 35
galactose 81
gelation of alginate 101–2, 137–9
D-glucosamine–nucleobase–peptide conjugates 40
glutamate–elastin-like protein conjugate 60
glutaraldehyde crosslinking 185
glycation of collagen 183
glycosaminoglycans *see* proteoglycans
gradient formation, temporal dynamics of 19–20
D-guluronic acid 136
α-L-guluronic acid (G) 97, 137, 139
    blocks 97, 98, 136
GVGVP pentapeptide–aromatic motif conjugates 34, 35

heart *see entries under* cardiac
HeLa cells, hydrogelator biocompatibility 35, 37–8, 39, 40
heparin 15–16, 50, 61
hepatocyte(s)
    cultures, in drug development/screening 198–9, 204
    phenotype, mesenchymal stem cell differentiation into 146
hepatocyte growth factor in myocardial regeneration 149
HGF (hepatocyte growth factor) in myocardial regeneration 149
Hippo 87
human amniotic membrane (HAM) 76, 77

hyaluronic acid/hyaluronan (HA) 81, 193–4
    gels 193–4
        cell therapy 200
        directed differentiation 196
        drug screening 204
        thiolated carboxymethyl (CMHA-S) 194, 196, 197, 200, 204
hydrogelators, small peptide, biocompatibility 31–47
hydrolytic degradation (hydrolysis) 20–1, 23
    alginate hydrogels 103, 136
    enzymatic degradation and, differences between 22
    enzyme-responsive hydrogels 119, 125, 128
hydroxyapatite–alginate scaffolds 155
hydroxymethyl cellulose with alginate gels 152

IGF-1 in myocardial regeneration 149
imaging, intracellular 126–8
immune response to hydrogelators 42–3
immunoisolatory properties of alginate gels 159
in situ
    cell environment, *see also* culture
    injectable *in situ* gelling alginate in myocardial regeneration 149
    sol to gel transition 56
in vitro
    cell environment 172
    tissue engineering 201–9
in vivo cell environment 172
indentation tests 174–6
induced pluripotent stem cells 74
inflation test 174

instructive role of biomaterials 81, 82
insulin-like growth factor 1 in myocardial regeneration 149
insulin-producing cells derived from embryonic stem cells 143, 160, *see also* islet (pancreatic) transplantation
integrin ligands/receptors 50, 83
internal gelation of alginate 10–32
interstitial convective flow 24, 25
intervertebral disc repair 155, 201
intracellular imaging 126–7
ionic strength hydrogels 23
islet (pancreatic) transplantation 155, 159

kidney injury, acute 201
kinetic control of biocatalytic peptide self-assembly 118–19
KTTKS sequence 81

β-lactamases in peptide self-assembly 117, 118
Lactobacillus 96
    microencapsulated in alginate gels 106
lactoferrin 14
laminin 128–9
laser-induced surface acoustic waves 176
LCD (liquid crystal display), enzyme-responsive hydrogel with 128
ligands, protein, extracellular 50
limbal epithelial cells 142, 152, 154
liquid crystal display, enzyme-responsive hydrogel with 128
liver-specific hydrogel 204, *see also* hepatocytes
Lustig–Peppas model 7, 21
Lys–Thr–Thr–Lys–Ser (KTTKS) sequence 81

magnetically-aligned collagen 186
MA–HA hydrogels 193–4, 196

mammalian cells, hydrogelator biocompatibility in 33–41
β-D-mannuronic acid (M) 97, 98, 137, 139
    blocks 97, 136
matrix (and matrices)
    controlled degradation 51–2, 113, 115
    extracellular *see* extracellular matrix
matrix metalloproteinases (MMPs) 52, 83–4, 118, 178
mechanical properties of hydrogels 171–90
    cellular activities influenced by 174–6, 181–4
    cellular activities influencing 178–81
    characterization 173–6
    strategies for improving 184–7
mechanoresponse 84
mechanosensing 84–5, 87
mechanotransduction 84, 86, 88
melanoma, malignant 128
mesenchymal stem cells (MSC) 144–6
    biochemical properties of alginate gels and 156
    biophysical and biomechanical properties influencing 177–8
        alginate gels 153
    differentiation
        directed 195–6
        to hepatocyte 146
    preclinical studies of therapeutic cell delivery 160
Metchnikoff, Ilya Ilyich 96
methacrylated HA (MA–HA) hydrogels 193–4, 196
microencapsulated alginate hydrogels *see* encapsulated/microencapsulated alginate hydrogels
microenvironments, tailorable 83, 197

Subject Index

microfluidic channels 12, 150
micro-indentation test 175
microparticle incorporation in hydrogels 13
micropatterning 186
mixing-induced two-component hydrogels 64-6
Mooney-Rivlin elasticity 140
morphology control in peptide self-assembly 118-22
MTT assays with hydrogelators 35, 37, 39, 40, 43
multidomain constructs 53-5
multipotent stem cells 74, 88, 150
mutagenesis, combinatorial 56
myocardial damage (incl. infarction) and its repair 75, 146, 148-9, 160, 200
myosin II, non-muscle (NMMII) 85, 86, 87

N-terminus unsubstituted dipeptides, biocompatibility 36-8
nano-indentation test 175-6
nanopatterning 186
nanostructures
    electrospinning of nanofibres 187
    self-assembling
        intracellular imaging 126-7
        peptide amphiphiles as 79-80
naphthalene (N)-containing hydrogelators 34, 35, 37
negatively charged molecules 13
neovascularization, therapeutic 149-50
network structure 183-4
    alginate gels 140, 141, 152
    indicators 183-4
    synthetic gel types distinguished by chemical nature of 113

neural stem cells 150-1
    biochemical properties of alginate gels and 156
    biophysical properties of alginate gels and 153
    preclinical studies of therapeutic cell delivery 162-3
niche, cell (bioartificial) 141, 142, 146, 151, 164, 172, 202
    recapitulation 197-8, 202
NMR spectroscopy, alginate 136
non-muscle myosin II (NMMII) 85, 86, 87
nuclear magnetic resonance spectroscopy, alginate 136
nucleobase-peptide conjugates, hydrogelators of 38-40
nucleobase-peptide-saccharide conjugates, hydrogelators of 40-1
nucleus pulposus, scaffold mimicking properties of 155

ocular therapy, alginate gels 142
ophthalmological (ocular) therapy, alginate gels 142
optic sensor for enzyme activity 128
optical coherence tomography 175
organoids, rotating wall vessel-generated 204
ovarian tissue culture, alginate hydrogels 199

pancreatic (islet) transplantation 155, 159
parathyroid tissue transplantation 159
PDLLA/bioglass 81
PEG (polyethylene glycol) hydrogel 8, 15, 16, 79, 83, 192, 192-3
    directed stem cell differentiation 195
PEGDA (polyethylene glycol diacrylate) hydrogels 192, 195, 199, 204

pentapeptides
  elastomeric 60, 61
  pentapeptide–aromatic motif conjugates 33–6
peptide(s), see also dipeptides; pentapeptides
  binding to soluble molecules 16–17, 27
  hydrogelators based on 116
    biocompatibility 31–47
    supramolecular 31–5, 41, 116
peptide amphiphiles 79–80
  aromatic 116, 121, 126
permeability (of hydrogels) 25–6, see also porous media
  alginate hydrogels 99, 141
pH-responsive hydrogels 23
pharmaceutical industry and cell-based therapy 75–6, see also drugs
phenotype, cell
  biochemical properties of alginate gels influencing 151–9
  biophysical and biomechanical properties influencing 177, 181
    alginate gel 151–9
phenylalanine–D-glucosamine–nucleobase conjugates 40
phosphatases in peptide self-assembly 117
plasmids containing engineered protein genes 56–7
pluripotent stem cells 74
  embryonic see embryonic stem cells
  induced 74
point-sinks 3, 8, 9, 11
point-sources 3, 8, 9
polyacrylamide hydrogels s, permeability 26
polyethylene glycol see PEG
poly(ethyleneimine) in alginate gels 140

poly-D-L-lactide/bioglass 81
poly-L-lysine (pLL)-coated alginate capsules/microspheres 107, 143, 149, 156
polymers/polymeric hydrogels 113
  concentration polymer concentration 185
  functionalization 13
porous media, see also permeability
  Brinkman equation/model for flow through 4, 5, 24
  porous alginate scaffolds 143, 153
positively charged molecules 13
preservation (cell), alginate gels in 142, see also cryopreservation
probiotic bacteria 96
  microencapsulated in alginate hydrogels 103–10
progenitor cells
  cardiac 149, 157, 163
  endothelial see endothelial cells
  stem cells and, definition and differences 74
proline-rich domains in mixing-induced two-component hydrogels 64–5
proteases in peptide self-assembly 117
protein
  ageing and structure of 183–4
  alginate microcapsules coated with 107
  ECM, influence on hydrogel properties 178–9
protein hydrogels, recombinant 48–72
proteoglycans and glycosaminoglycans (GAGs) 13, 15, 17
  sulfated 83, 179
proteolysis 14, 51–2
pullulan hydrogel 201
pyrene (P)-conjugated gelator peptide 34, 35, 36

Subject Index

reaction–diffusion model 17–19
recombinant protein hydrogels 48–72
regenerative medicine 194–208
    biomaterials 88
    cornea 141–2
    ECM-like hydrogels 194–208
renal injury, acute 201
repulsion (electrostatic), ECM–solute 13
resilin-like proteins 63–4
responsive hydrogels 23–4
    enzyme 112–34
Reynolds numbers 12
RGD (arginine–glycine–aspartic acid) sequence 80, 83, 113, 124
    alginate hydrogels 143, 148, 149, 155, 156, 157, 158, 159
    PEG gels 195
RGDS (arginine–glycine–aspartic acid–serine) sequence 81, 83
    alginate gels, and embryonic stem cells 143, 156
    PEG gels 195
Rho 87
ROCK (Rho-associated protein kinase) 87
rotating wall vessel (RWV) technology 202–4

saccharide–nucleobase–peptide conjugates, hydrogelators of 38–40
scaffold(s) tissue/cell 78, 82, 128–9
    alginate hydrogels, for cell culture 141–51
    enzyme-responsive hydrogels 128–9
    stem cells and 75–6
    tuneable 80, 82
scaffold-free bioprinting 207
Schmidt numbers 12
self-assembling hydrogels *see* supramolecular hydrogels
self-assembling peptide amphiphiles 79–80

sequence (peptidic), morphology control by 121–2
smart biomaterials 79, 88, 114–15
soluble molecules/factors 1–30
    alginate microcapsules coated with 107
    generation and consumption
        in ECM 8–11
        in hydrogels 11–13
    transport in hydrogels vs ECM 1–30
        temporal dependencies 17–24
spherical indentation test 175
spongy (trabecular) bone engineering 144–8
steady state diffusion *see* diffusion
stem cells (for therapy/transplantation etc.) 67, 73–4, 75–6, 195–8
    adipose-derived *see* adipose-derived stem cells
    adult (in general) *see* adult stem cells
    amniotic fluid 198
    banking 76
    biomechanical properties influencing 177–8
    bone marrow (bovine) 159
    clinical applications 75–6
    in corneal regeneration 141–2
    cryopreservation 76, 143–4
    cultures 195–8
    embryonic *see* embryonic stem cells
    mesenchymal *see* mesenchymal stem cells
    multipotent 74, 88, 150
    neural *see* neural stem cells
    pharmaceutical industry and 75–6
stiffness 65, 82, 177–8
    cellular activity affecting 179–80
    niche recapitulation and 197
    strategies targeting 184, 185
Stokes–Einstein relationship 4, 5, 6

streptococcal collagen-like protein 63
strip extensiometry 173–4
strontium gels 139, 148
substrate, elasticity, cell response to 81–3
subtilisin 118, 119, 122, 128
sulfated glycosaminoglycan 83, 179
supramolecular (self-assembled) hydrogels 31–5, 41, 113, 115–25
    biocatalytic assembly 115–24
surface acoustic waves, laser-induced 176

T98G cell lines, hydrogelator biocompatibility 37–8
tailorable microenvironments 83, 197
TAZ (transcriptional co-activator with PDZ-binding motif) 87
temperature-responsive hydrogels 23–4
temporal (time) aspects
    ability of hydrogels to dynamically change over time 209
    soluble molecule transport in ECM vs hydrogels 17–24
tensile testing 173–4
tensioning-culture force monitor system 176
TGF *see* transforming growth factor
thermodynamic control of biocatalytic peptide self-assembly 120–2
thermolysin 120, 122, 128
thermoresponsive hydrogels 23–4
thiolated carboxymethyl HA (CMHA-S) gels 194, 196, 197, 200, 204
three-dimensional (3D) cell culture 171
    drug screening 202–4
    orientation of fibrous components 186
    using alginate hydrogels 141–51
three-dimensional (3D) tissue printing/bioprinting 149, 205–8
time *see* temporal aspects
tissue culture *see* culture
tissue engineering
    biomaterials in 78–88
    bone and cartilage 144–8
    cardiac 149
    enzyme-responsive hydrogels 128–9
    in vitro and *ex vivo* 201–9
tissue printing/bioprinting 149, 205–8
trabecular bone engineering 144–8
transcriptional co-activator with PDZ-binding motif (TAZ) 87
transferrin 14
transforming growth factor (TGF)
    binding sequences in biomaterial 82
    scaffold for TGF-β1 presentation and sustained release 159
transplantation 159
    endothelial progenitor cells 150
    in ocular therapy 142
    pancreatic islets 155, 159
    parathyroid tissue 159
    stem cells *see* stem cells
transport of soluble molecules *see* soluble molecules
tropoelastin 58–9, 60, 61
tuneable substrates 88
    scaffolds 80, 82
    tuneable *see* tuneable substrates
two-component hydrogels, mixing-induced 64–6
tyrosinase and malignant melanoma 128

ultrasound elastography 176

vascular endothelial growth factor (VEGF) 13, 16, 162
VEGF (vascular endothelial growth factor) 13, 16, 162
viability, cell, mechanical properties of gels as indicators of 182–3
viscoelastic properties
    alginate hydrogels 140–1
    assessment 173, 174, 175
VTEEI pentapeptide–aromatic motif conjugates 34, 35, 36
VYGGG pentapeptide–aromatic motif conjugates 34, 35

weight of animals, hydrogelators effects on 43
wound healing assays, hydrogelators 43

xanthan–alginate microspheres 109

YAP 85, 87–8
Yes-associated protein (YAP) 85, 87–8
YGFGG pentapeptide–aromatic motif conjugates 34, 35

zebrafish, laminin-deficient 129
Zener model (viscoelasticity) 141